Construction Economics and Cost Management for Civil Engineers

Construction Economics and Cost Management for Civil Engineers

Edited by **Sarah Crowe**

WILLFORD PRESS

New York

Published by Willford Press,
118-35 Queens Blvd., Suite 400,
Forest Hills, NY 11375, USA
www.willfordpress.com

Construction Economics and Cost Management for Civil Engineers
Edited by Sarah Crowe

International Standard Book Number: 978-1-68285-015-2 (Hardback)

Printed in the United States of America.

Contents

Preface

This book is a compilation of chapters that discuss the most vital concepts and emerging trends in the field of civil engineering. Thoroughly elucidated in this book are significant concepts of construction economics, such as quantity surveying, property management, etc. It is compiled in such a manner, that it will provide in-depth knowledge about the various theories and their applications for construction economics procedures. The extensive content of this book will provide the readers with a comprehensive understanding of the emerging topics and trends of this subject.

This book is a comprehensive compilation of works of different researchers from varied parts of the world. It includes valuable experiences of the researchers with the sole objective of providing the readers (learners) with a proper knowledge of the concerned field. This book will be beneficial in evoking inspiration and enhancing the knowledge of the interested readers.

In the end, I would like to extend my heartiest thanks to the authors who worked with great determination on their chapters. I also appreciate the publisher's support in the course of the book. I would also like to deeply acknowledge my family who stood by me as a source of inspiration during the project.

Editor

Performance Measurement in the UK Construction Industry and its Role in Supporting the Application of Lean Construction Concepts

Saad Sarhan and Andrew Fox, (University of Plymouth, UK)

Abstract

Performance measurement has received substantial attention from researchers and the construction industry over the past two decades. This study sought to assess UK practitioners' awareness of the importance of the use of appropriate performance measures and its role in supporting the application of Lean Construction (LC) concepts. To enable the study to achieve its objectives, a review of a range of measurements developed to evaluate project performance including those devoted to support LC efforts was conducted. subsequently a questionnaire survey was developed and sent to 198 professionals in the UK construction industry as well as a small sample of academics with an interest in LC. Results indicated that although practitioners recognise the importance of the selection of non-financial performance measures, it has not been properly and widely implemented. The study identified the most common techniques used by UK construction organisations for performance measurement, and ranked a number of non-financial key performance indicators as significant. Some professed to have embraced the Last Planner System methodology as a means for performance measurement and organisational learning, while further questioning suggested otherwise. It was also suggested that substance thinking amongst professionals could be a significant hidden barrier that militates against the successful implementation of LC.

Keywords: Construction Industry, Lean Construction, Project Performance Measurement, Psychology, UK

Introduction

A growing number of companies worldwide began to recognise the benefits that could be achieved from adopting the lean construction (LC) approach (Arbulu and Zabelle, 2006). Many practitioners and academics have reported case studies where their companies were achieving some proven benefits and pockets of excellence (Alarcon et al., 2002; Swain and Mossman, 2003; Wu and Low, 2011; Andersen et al., 2012; Keiser, 2012). However, a number of studies in countries across the globe have revealed that the application of lean principles to construction has not been successful due to a number of critical factors/barriers (Olatunji, 2008; Senaratne and Wijesiri, 2008; Abdullah et al., 2009; Mossman, 2009a). One of these factors is the failure to use appropriate process performance measurement systems (PMS); which is crucial to support the implementation of LC (Sarhan and Fox, 2012).

Although process performance measurements have received substantial attention from academic researchers over the past two decades, the construction industry still has a preference for measureing performance in terms of time and cost (Bowen et al., 2002; Forbes et al., 2002). These traditional (results-based) performance preferences measured in projects, specifically costs and schedule, are not appropriate for continuous improvement because they are not effective in identifying the root-causes of quality and productivity losses (Alarcon and Serpell, 1996). Despite this fact, there are very few, if any, studies that have sought to explore and understand the reasons why professionals and managers in the construction industry still have a tendency to measure performance in terms of time, cost and meeting code, as opposed to process performance measures (e.g. cycle time, Rework, waste). A series of ontological work by Rooke et al. (2003; 2004; 2007) strongly suggests that the overreliance of managers and professionals on objects rather than processes as the key to understanding and communicating about construction projects, presents a barrier to learning flow-based solutions. Based on a review of a range of measurements developed to evaluate project performance including those devoted to support lean construction efforts, as well as a systematic consideration to the series of work done by Rooke et al (2003; 2004; 2007), this research carried out a survey among UK professionals to assess their awareness of the importance of the use/selection of appropriate performance measures, crucial to support the implementation of the LC approach.

Lean Thinking in Construction

Lean thinking is a philosophy based on the concepts of lean production. The first consideration of the ideas of lean production for use within construction is attributed to Koskela (1992) (Garnett et al., 1998; Mossman, 2009a). Koskela (1992) formulated the transformation-flow-value generation model of production, known as the TFV theory of production, which could lead to improved performance when applied to construction. This seminal technical-report (Koskela, 1992) proposed the need to review construction production as a combination of conversion and flow processes to remove waste, when traditional thinking of construction was only focusing on conversion activities and ignoring flow and value considerations (Garnett et al., 1998; Senaratne and Wijesiri, 2008). There are eight types of waste which are commonly agreed on by researchers: Transportation, Inventory, Motion, Waiting, Over-Production, Over-Processing, Defects, Skills Misuse (Terry and Smith, 2011). Consequently, many researchers emphasised the importance of the use of appropriate performance measurement systems, which can give early warnings and identify problems before they occur, to support the successful implementation of lean construction (Lantelme and Formoso, 2000; Alarcón et al., 2001; Leong and Tilley, 2008; Sarhan and Fox, 2012).

Performance Measurement Systems

The use of simple and well-designed performance measurement systems (PMSs) is essential for supporting the implementation of business strategies, such as the application of LC concepts within construction organisations. That is because performance measurement provides the information required for process control and makes it possible to set up challenging goals (Lantelme and Formoso, 2000; Moon et al., 2007). Neely et al. (1996) define performance measurement as *"the process of quantifying effectiveness and efficiency of action "*. Effectiveness is the extent to which a target is achieved (e.g. client satisfaction) with resources applied (Neely et al., 1996; Cheng et al., 2009). Efficiency is the evaluation of how economically the resources are utilised to meet client requirements (Neely et al., 1996).

Without the use of appropriate PMSs, it becomes very difficult for organizations to understand why poor performance continues, or how improvement could be achieved (Leong and Tilley, 2008). In addition, without PMSs managers cannot know whether they will be able to achieve their intended objectives and goals or not (Neely et al., 1996). According to Chrysostomou (2000) *"to manage you must measure, if you don't you are only practising"*; cited in Alarcón et al. (2001). This points out that the selection of appropriate measures has a major influence on the implementation of strategies, and is essential for the development of improvement programmes (Lantelme and Formoso, 2000).

Result-Based Indicators vs. Process-Based Indicators

Traditional performance measurement systems are based on financial measures (Lantelme and Formoso, 2000; Suwignjo et al., 2000). These financial measures are result-oriented performance indicators, and have been strongly criticised by many researchers (Alarcón et al., 2001; Mitropoulos and Howell, 2001; Takim and Akintoye 2002; Costa et al. 2004; Moon et al., 2007; Nudurupati et al., 2007; Leong and Tilley, 2008). That is because these parameters are backward focused (Lantelme and Formoso, 2000). They are not measured until project is complete; and thus the information obtained arrives too late to take any corrective actions (Alarcón and Serpell, 1996; Moon et al., 2007). As a result, these outcome based indicators cannot be used to identify barriers or problems that exist during the execution of processes. According to Alarcón et al. (2001) traditional control systems focus their attention in conversion activities and ignore flow activities; therefore nearly all non-value-adding activities become invisible.

Instead, Costa et al. (2004) recommend the use of (process-oriented) leading measures aiming to give early warnings, identify barriers and potential problems, and emphasize the need for future investigation. This recommendation is supported by Neely et al. (1996) who asserted the need to adopt formal process based approaches. It is important to use measures for tracking improvement not reporting (Terry and Smith, 2011). Likewise, Alarcón et al. (2001) emphasise that measurement alone is not enough; it is essential to analyse these indicators with the objective to detect the problems and their root causes.

Key Performance Indicators (KPIs) in the UK

There is an industry tendency to measure performance in terms of time, cost and meeting code; but very limited consideration has been subjected to client satisfaction (Forbes et al., 2002). In order to help organisations move towards best practice in response to the Egan's report (1998), the UK working groups on KPIs identified a set of non-financial parameters for benchmarking projects (Takim and Akintoye, 2002; Dawood et al., 2006). These KPIs are classified into three levels, namely, headline, operational, and diagnostic (Costa et al., 2004; Nudurupati et al., 2007). Headline indicators provide a measure of the overall health of a firm. Operational Indicators bear on specific aspects of a firm's activities and should enable management to identify and focus on specific areas for improvement. Diagnostic Indicators provide information on why certain changes may have occurred in the headline or operational indicators and are useful in analysing areas for improvement in more detail (The KPI Working Group, 2000).

Despite the KPI programme, there are some problems identified in the KPIs. For instance, none of the measures mentioned could identify the performance of suppliers in a project environment (Takim and Akintoye, 2002; Costa et al., 2004). Also, there are no suggestions for performance indicators in benchmarking projects at the project selection phase, such as the analysis stage (Takim and Akintoye, 2002). For this reason, Takim and Akintoye (2002) propose that the successful construction project performance can be divided along three orientations: procurement, process, and results oriented. A similar approach was adopted by Sikka et al. (2006) who classified KPIs into three conceptual phases of a construction project: pre-construction, construction, and post-construction; as they believe that project success criteria change with time in each phase.

The Balanced Scorecard

The balanced scorecard (BSC) is a widely accepted framework (Nudurupati et al., 2007). It was constructed to complement measures of past performance with measures of the drivers of future performance (Nudurupati et al., 2007). It links an organisation's strategy through a series of perspectives to KPIs (Fraser and Kelly, 2011). According to Karanseh and Al-Dahir (2012) the BSC performance measurement model as presented by Kaplan and Norton (1992) is a business management concept that is more focussed on strategy and vision rather than control. However, it could be argued that it can be difficult and confusing to integrate between the BSC's strategic and operational level measures (Karanseh and Al-Dahir, 2012).

Quantitative Models for Performance Measurement Systems (QMPMS)

QMPMSs use cognitive maps, cause and effect diagrams, tree diagrams, and the analytic hierarchy process, to quantify the effect of factors on performance (Suwignjo et al., 2000; Nudurupati et al., 2007). There are three main steps in QMPMS: (1) identifying the factors that affect performance and their relationships; (2) structuring the factors hierarchically; (3) quantifying the effect of factors on performance (Suwignjo et al., 2000; Nudurupati et al., 2007). The quantification process is carried out based on the results of a pair-wise comparison questionnaire among the factors (subjective technique).

This approach for quantifying the effects of factors on performance could be criticised because it is subjective, and it may be difficult to be applied in practices. One of the potential problems of this approach is that performance improvement usually involves identification of a large number of factors affecting performance. Consequently, the number of pairwise comparison questionnaire will be huge; and filling it in will be exhausting and time consuming (Suwignjo et al., 2000).

The Last Planner System and Lean-based Process Measures

The Last Planner System (LPS) for production control (Ballard, 2000) has been implemented in construction projects with varying levels of success, to increase the reliability of planning, improve production performance, and create a predictable workflow (Hamzeh et al., 2009). Through the LPS methodology, project teams commit to complete assigned tasks in a given week. Some LC practitioners refer to percentage plans complete (PPC) as a metric for commitment reliability. According to Forbes and Ahmed (2011) a PPC value does not measure the level of utilization of a work flow (efficiency). Instead it measures production planning effectiveness and workflow reliability. At each weekly meeting, time is given to learn and understand why certain tasks were not completed as planned in the previous week, before creating a new weekly plan to be executed. The uncompleted plans are studied and analysed to determine the barriers and root causes that affected the implementation process. The five-WHY analysis procedure could be used for identifying the root-causes of problems; and a Pareto chart could be used for ranking the barriers and reasons for non-completion. Consequently, the information gained from the root-cause analysis would help the project teams to avoid obstacles in future work cycles, and improve the effectiveness and reliability of future work plans (Forbes and Ahmed, 2011).

Many researchers, over the last two decades, have published papers which included a wide range of process and lean-oriented performance measures for use within the constrction industry. For example, Alarcón et al. (2001) suggested a set of parameters that are lean based, and which could help companies to measure waste, cycle times and re-work in construction projects. Moon et al. (2007) proposed a set of process-oriented performance indicators which are derived from the TFV theory: reliability, efficiency and effectiveness. Leong and Tilley (2008) proposed a lean strategy to performance measurement, which aims to reduce waste in projects by measuring next customer needs. In general, the essence in these approaches is to create a measurement culture, within organisations, that will facilitate future implementations. More lean performance measures can also be found at Forbes and Ahmed (2011). However, it seems that some of these initiative lean performance measures still require further development and experimentation, in order to gain wide acceptance in the construction industry.

Predominance of Results-Based Solutions in the Thinking of Professionals in Construction

Ontological work by Rooke et al. (2003; 2004; 2007) emphasised that the most successful production management solutions are flow based ones; and that adherence to substance (results based) thinking poses a significant barrier to achieving progress in the construction industry. This suggestion is strongly supported by educational psychology work by (Itza-Ortiz et al., 2003) where it was observed that students in general tend to face difficulties in absorbing process-based theories, in contrast to more simply understood substance-based ones.

An example of this in the UK is the use of bills of quantity (BoQ) based on the Civil Engineering Standard Method of Measurement (CESSM). That is because there are two problems, which could lead to price variations and delivery difficulties, that exist with CESSM based bills. First, aggregating the BoQ items into self-contained construction operations is done by client representives and may not match the way the contractor intends to do the works (Hoare and Broome, 2001). The second is due to the lack of transparency in the way that prices are made up as to the contractor's assumptions about profit and quality of work (Rooke et al., 2007). In total, two methods were identified by Rooke et al. (2007) as examples of results-based thinking (CESSM and the design/construction dichotomy) while three as examples of flow-based thinking (the activity schedule, the LPS and claims planning).

Research Method

This research study aims to assess UK practitioners' understanding of the importance of the use of appropriate PMSs and its role in supporting the implementation of initiative business strategies such as lean construction. In order to achieve this aim, the following research objectives were derived:

- Conduct a literature review of a range of measurements developed to evaluate project performance including those devoted to support lean construction efforts;
- Identify the most common techniques used by construction organisations for performance measurement;
- Identify and rank the significant (most important) non-financial key performance indicators for construction organisations;
- Identify how LPS is being used within construction organisations in terms of its areas of application;
- Assess practitioners' understanding to the function of the PPC value within the LPS.

An invitation to complete a questionnaire survey was sent to 198 professional practitioners in the UK construction industry as well as a small sample of academics with an interest in LC (10 for a pilot study and 188 for the main study). Participants were selected randomly from a number of professional groups that represent most of the professional organisations involved in the UK construction industry. These groups are the Chartered Institute of Building (CIOB), Institution of Civil Engineers (ICE), Royal Institute of British Architects (RIBA), Institute of Highway Engineers (IHE), Royal Institution of Chartered Surveyors (RICS), and the Lean Construction Institution in the UK (LCI-UK). Academics formed 4% of the sample and were selected incidentally as they hold professional status and qualifications. However, their responses were not excluded from the analysis of the survey because they specialise in construction management and have wide industry experience. A total of 140 responses were received representing a response rate of 74.5% (SE=0.032). Some techniques were adopted to help the study to capture this high rate of response. These include:

- Pilot studies - Piloting is good research practice and is part of the research Plan-Do-Check-Act process (Lancaster et al., 2004).
- Both the invitation letter as well as the front cover of the questionnaire included an information sheet for the participants, which illustrated clearly what the research was about, what it involved, and ensured data protection.
- Follow ups and reminders by direct contact. Each participant of the sample was contacted individually by email.

The distribution/mixture of the professionals and organisations involved in this study are shown in Tables 1 and 2. The largest proportion of the participants was for civil engineers (34%). In addition, more than half of the respondents (63%) were from practitioners holding managerial positions and with more than 10 years of experience in the industry.

[1] Years of experience			[2] Current role (Managerial level)				[3] Level of education		
0-10	10-20	20+	Graduate /Junior	Middle management	Senior management	Other	Practical qualification	Bachelor's degree	Master's Degree and above
37%	26%	37%	14%	26%	22%	37%	25%	36%	39%

Table 1 Distribution of the sample in percentage (Clustering of individuals)

[1] Average Annual Turnover in £ Millions			[2] Size of organisations		[3] Major Client		
1-100	100-1000	1000+	<500 employees	>500 employees	Private	Public	Both
40%	31%	29%	46%	54%	14%	26%	60%

Table 2 Distribution of the sample in percentage (Clustering of organisations)

Research Results and Analysis
Techniques used by construction organisations for performance measurement
A range of measurements developed to evaluate project performance were identified from literature and introduced to a question, in order to allow the study to identify the most common techniques used by construction organisations for performance measurement. The responses revealed that result-oriented KPIs are the most common technique used amongst construction organisations for performance measurement. It was also found that many organisations still rely heavily on the experience of their managers as means for performance measurement (Figure 1).

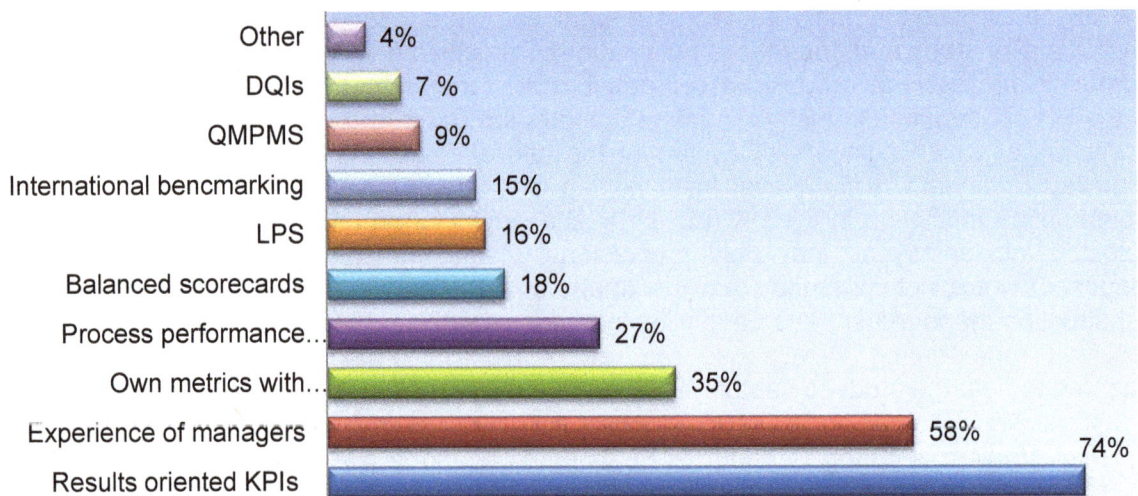

Technique	Percentage
Other	4%
DQIs	7%
QMPMS	9%
International bencmarking	15%
LPS	16%
Balanced scorecards	18%
Process performance...	27%
Own metrics with...	35%
Experience of managers	58%
Results oriented KPIs	74%

Figure 1 Techniques used by construction organisations for performance measurement

Significant Key Non-financial Performance Measures

A question was introduced to rank and identify the significant 'non-financial' key performance indicators (leading indicators) that are appropriate for continuous improvement, according to their importance to organisations. The respondents were asked to rate the given non-financial performance measures on a ten-point scale to indicate the level of importance (10 being the most important and 1 being the least important). A 10-point, end-defined, scale was adopted for this question based on feedback received from pilot studies because it gives more room to the respondents for a real appreciation with the 10 possibilities. According to Cummins and Gullone (2000) *"A review of the literature indicates that expanding the number of choice-points beyond 5- or 7-points does not systematically damage scale reliability, yet such an increase does increase scale sensitivity"*.

The mean values of the given non-financial performance measures were then determined to indicate the degree of importance of these performance measures to construction organisations from the perspective of the respondents (Table 3). If the mean value scored ''8'' or above to a particular performance measure, then it would be classified as a significant performance measure. In similar research, Cheng et al.(2001), Chan et al. (2003), and Lam et al. (2007) represented the level of significance on a five-point Likert scale by a score of ''4''.

Rank	Non-financial performance measures	Mean value score out of 10
1	Safety	9.504
2	Client/customer satisfaction	9.149
3	Quality	8.775
4	Team Performance	7.803
5	Productivity	7.785
6	Functionality	7.654
7	Planning Efficiency	7.607

Table 3 Ranking of non-financial key performance indicators according to their importance to construction organisations

Note: The shaded areas represent the significant performance measures

Table 3 shows that the mean values of safety, client satisfaction and quality exceed the cut-off point (a score of 8), and thus are considered as the significant (most important) non-financial KPIs for construction organisations. A reliability test was conducted for this question and Cronbach's Alpha value was found to be 0.832; which indicates a high degree of reliability, as a value ≥ 0.7 is considered to be acceptable (Lam et al., 2007; Ab Rahman et al. 2011).

Areas of Application of LPS within Construction Organisations

A question was then added to identify how LPS is being used within organisations. Also, to determine whether organisations are aware of the full benefits of LPS and its importance in providing means for performance measurement and organisational learning, or if LPS is just seen by them as an activity scheduling tool. The respondents had the chance to choose more than one answer. More than half (61%) of the respondents stated that the question is 'Not Applicable', while 39% of them acknowledged that LPS is used by their organisations. Figure 2 below shows the arrangement of the different areas of application of LPS, according to their frequency of use within organisations involved in this study.

LPS is used in your organisation for:

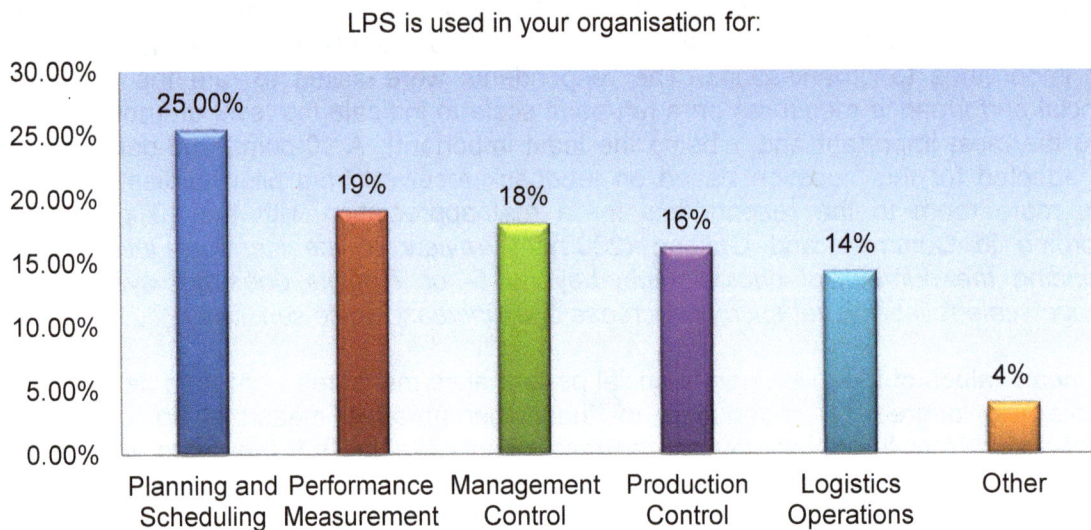

Figure 2 Areas of application of LPS within construction organisations

Practitioners' Understanding to the Function of the PPC value within the LPS

Respondents were then asked if the Percentage Plan Complete (PPC) value in LPS: (a) measures the level of utilisation of work flow (efficiency), (b) measures production planning effectiveness and work flow reliability, or (c) both of [a] and [b]. The aim of this question was to examine the respondents' awareness/understanding to the function of the PPC value when using the LPS, as a means for performance measurement. More than half of the respondents (62.5%) mentioned that the question is not applicable to them; and only 15% of all respondents were able to answer the question correctly.

Discussion

The selection of appropriate measures has a major influence on the implementation of strategies, and is essential for the continuous development of improvement programmes (Lantelme and Formoso, 2000). That is because without the use of appropriate PMSs, it becomes very difficult for organizations to understand why poor performance continues or how improvement could be achieved (Leong and Tilley, 2008). Based on a review of a range of measurements developed to evaluate project performance, a number of process-oriented (leading) measures were selected and provided alongside other traditional (lagging) measures; to determine the techniques used by construction organisations for performance measurement. The results revealed that professionals rely heavily on results-based KPIs as opposed to process performance measures. It was also found that many organisations still depend on the experience of their managers as a means for project performance measurement.

Despite the fact that continuous improvement requires analysis of processes and devising an internal metrics for evaluating performance, only about 35% of the respondents mentioned that their organisations use their own metrics that is consistent with their business strategy. Also, process (non-financial) performance measures and LPS, which are recommended by researchers for providing means to support the successful implementation of LC, were selected by just 27% and 16% of all respondents respectively. However, when the respondents were asked to give a score out of ten (10 being the most important and 1 being the least important) to some non-financial performance measures, the mean values of three measures namely: safety, client satisfaction and quality obtained a score above 8 (most significant); and the lowest score obtained amongst all other performance measures was 7.6. These results indicated that although construction practitioners recognise the importance of the use of process and non-financial performance measures, it has not been properly and widely implemented in the construction industry. Most managers still make decisions just based on their experience and common sense, and on a few traditional financial measures which are no longer suitable in the existing competitive environment. These two approaches, in particular, are no longer appropriate for continuous improvement, and have been strongly criticised by many researchers.

The Last Planner System (LPS) is a production planning and control system based on lean production principles, which is developed to improve planning reliability and project performance (Gonzalez et al., 2010). It introduces the next customer into the equation through continuous and collaborative planning (Johansen and Walter, 2007; Mossman, 2009b). Through the LPS methodology, project teams commit to complete assigned tasks in a given week using functions such as: 'Look-Ahead Plan' to plan what can be done when constraints are removed, and the Percent Planned Complete (PPC) which monitors the Look-ahead Plan and requires reasons for delays which are analysed in terms of root causes (Ansell et al., 2007). Some LC practitioners refer to percentage plans complete (PPC) as a metric for commitment reliability. From results obtained, it appeared that LPS is not widely used among construction organisations. Less than half of the respondents (39%) acknowledged that LPS is being used within their organisations; and the majority of these respondents were not aware of the full benefits of LPS and its importance in providing means for performance measurement and organisational learning. Instead, it is mainly employed by their organisations just as an activity scheduling tool. These results are consistent with the findings of Common et al. (2000).

In general there seemed to be inconsistencies in the answers of the respondents regarding how LPS is employed within their organisations, especially when it comes to its application as a means/tool for performance measurement and organisational learning (See Figures 1 and 2). Furthermore, it was found that a minority of all respondents (only 15%) were aware of what the PPC ratio actually measures. This contradiction suggests that there is a considerable lack of awareness of the application of LPS and its role, in terms of creating conditions for decentralised control and providing a learning process at operational level (Lantelme and Formoso, 2000).

Ontological work by Rooke et al. (2003; 2004; 2007) emphasised that the most successful production management solutions are flow based ones; and that adherence to substance (results based) thinking poses a significant barrier to achieving progress in the construction industry. Thus, it could be argued here that if the overreliance of the respondents/practitioners on using results based KPIs would be classified as an example of substance (object) thinking as opposed to process performance measures - e.g. of process thinking, then accordingly it is suggested that substance thinking amongst professionals acts as a significant *hidden* barrier that mitigates against the successful implementation of LC. The term *hidden* is used by the authors to describe this sort of barrier because it has not yet been empirically demonstrated in construction research. However, this suggestion is strongly supported by a series of ontological work conducted by Rooke et al. (2003; 2004; 2007); and is linked to educational psychology work by (Itza-Ortiz et al., 2003) where it was observed that students in general tend to face difficulties in absorbing process-based theories, in contrast to more simply understood substance-based ones. Academic researchers are highly recommended to explore the root causes of this problem (adherence of practitioners to results-based solutions). One way to address this issue could possibly be achieved through investigating the role that psychology can play in the educational framework, in terms of facilitating the learning process of flow based concepts and solutions amongst students, within the construction and built environment.

Conclusions and Recommendations

Lean construction efforts could prove to be highly rewarding for the construction industry, and could possibly lead the UK construction industry's quest to improve quality, efficiency and customer satisfaction. Several studies emphasised that the selection of appropriate measures has a major influence on the implementation and development of initiative business strategies (i.e. lean construction). Results of this study revealed that professionals rely heavily on results-based Key Performance Indicators (KPIs) as opposed to process performance measures. It was also found that many organisations still depend on the experience of their managers as a means for project performance measurement. These two approaches are not suitable for supporting the application of LC concepts and have been criticised by many researchers because they are no longer appropriate for continuous improvement.

Although, construction practitioners recognise the importance of the use of process and non-financial performance measures, it has not been properly and widely implemented in the UK construction industry. However, three non-financial KPIs were determined by respondents as highly important for construction organisations, namely: safety, client satisfaction and quality respectively. Hence it is important to link academic efforts with industry needs; therefore researchers are encouraged to focus their efforts on developing these three non-financial KPIs identified by the study so they can be applied by organisations and their supply-chain throughout all stages of construction projects.

The study indicated that there is a limited application of Last Planner System (LPS) among UK construction organisations. Some of the respondents professed to have embraced the LPS methodology as a means for performance measurement and organisational learning; while further questioning suggested that there is a considerable lack of awareness of the application of LPS and its role, in terms of creating conditions for continuous improvement and providing a learning process at operational level.

There are very few, if any, studies that have sought to explore and understand the underlying causes of why professionals in the construction industry adhere to results-based solutions, as opposed to flow-based ones. This study identified the overreliance of the practitioners on using results-based KPIs as an example of substance (object) thinking as opposed to process performance measures - e.g. of process thinking. Accordingly, it is suggested that substance thinking amongst professionals could be considered a significant *hidden* barrier that militates against the successful implementation of LC. This suggestion is linked to the series of ontological work by Rooke et al. (2003; 2004; 2007) and supported by educational psychology work by Itza-Ortiz et al. (2003). However, further investigation and experiential research is recommended to validate this suggestion.

Limitation

This study included many of the UK's largest construction organisations, while it could be argued that most of the construction companies in the UK are micro and small. Therefore, the authors suggest that a larger and more random sample is required to generalise and validate the findings of the study.

References

Ab Rahman, M. N., Shokshok, M. A. and Abd Wahab, D. (2011) 'Barriers and benefits of total quality management Implementation in Libyan manufacturing companies', *Middle-East Journal of Scientific Research*, **7** (4), 619-624

Abdullah, S., Abdul-Razak, A., Abubakar, A. and Mohammad, I. S. (2009) 'Towards producing best practice in the Malaysian construction industry: The Barriers in implementing the lean construction approach', Faulty of Engineering and Geoinformation science, Universiti Teknologi, Malaysia

Alarcon, L. F. and Serpell, A. (1996) 'Performance measuring, benchmarking, and modelling of project performance', *Proceedings for the 5th International Conference of the International Group for Lean Construction (IGLC-5)*, The University of Birmingham, UK

Alarcón, L., Grillo, A., Freire, J. and Diethelm, S. (2001) 'Learning from collaborative benchmarking in the construction industry', *Proceedings for the 9th International Conference of the International Group for Lean Construction (IGLC-9)*, Singapore

Alarcon, L. F., Diethelm, S. and Rojo, O. (2002) 'Collaborative implementation of lean planning systems in Chilean construction companies', *Proceedings for the 10th Annual Conference of the International Group for Lean Construction*, Gramado, Brazil, 6 - 8 August 2002

Andersen, B., Belay, A. M., and Amdahl Seim, E. (2012) 'Lean construction practices and its effects: A case study at St Olav's Integrated Hospital, Norway', *Lean Construction Journal*, 2012, 122-149

Arbulu, R., and Zabelle, T. (2006) 'Implementing Lean in Construction: How to Succeed', *Proceedings for IGLC-14*, Santiago, Chile

Bowen, P. A., Cattel, K. S., Hall, K. A., Edwards P. J., Pearl, R. G., (2002) *'Perceptions of Time, Cost and Quality Management on Building Projects'*, *Australasian Journal of Construction Economics and Building*, **2** (2), 48-56

Chan, A. P. C., Chan, D.W.M., Chiang, Y.H., Tang, B.S., Chan, E.H.W. and Ho, K.S.K. (2003) 'Exploring critical success factors for partnering in construction projects', *Journal of Construction Engineering and Management*, **130** (2), 188-198

Cheng, M., Tsai, H. and Lai, Y. (2009) 'Construction management process reengineering performance measurements', *Automation in Construction*, **18,** 183-193

Costa, D., Formoso, C., Kagioglou, M. and Alarcón, L. (2004) 'Performance measurement systems for benchmarking in the construction industry', *Proceedings for IGLC-12*, Copenhagen, Denmark

Cummins, R.A. and Gullone, E. (2000) 'Why we should not use 5-point Likert scales: The case for subjective quality of life measurement', *Proceedings for Second International Conference on Quality of Life in Cities* (pp.74-93), Singapore: National University of Singapore

Dawood, N, Sikka, S, Marasini, R and Dean, J (2006) 'Development of key performance indicators to establish the benefits of 4D planning', *In:* Boyd, D (Ed) *Proceedings 22nd Annual ARCOM Conference*, 4-6 September 2006, Birmingham, UK, Association of Researchers in Construction Management, 709-718

Egan, J. (1998) *Rethinking Construction: Report of the Construction Task Force*, London: HMSO

Fraser, N. and Kelly, R. (2011) 'Applying a balanced score card approach to waste reduction KPIs in Lean construction', [Online] Available at: www.ciria.org (Accessed: July 2011)

Forbes, L. H., Ahmed, S. M. and Barcala, M. (2002) 'Adapting lean construction theory for practical application in developing countries', *Proceedings of the first CIB W107 International Conference: Creating a Sustainable Construction Industry in Developing Countries*, Stellenbosch, South Africa, 11-13 November

Forbes, L. H. and Ahmed, S. M. (2011) *Modern Construction: Lean Project Delivery and Integrated Practices.* London: CRC Press

Garnett, N., Jones, D. T. and Murray, S. (1998) 'Strategic application of lean thinking', *Proceedings IGLC 98*, Guaruja, Brazil

Gonzalez, V., Alarcon, L., Maturana, S., Mundaca, F., and Bustamante, J. (2010) 'Improving planning reliability and project performance using the reliable commitment model', *Journal of Construction Engineering and Management*, **136** (10), 1129-1139

Hamzeh, F., Ballard, G., and Tommelien, I. (2009) 'Is the last planner system applicable to design? A case study', *Proceedings of the 17th Annual Conference of the International Group for Lean Construction*, IGLC 17, 13 -19 July, Taipei, Taiwan, 165-176

Itza-Ortiz, S.F., Rebello, S. and Zollman, D. (2003) 'Students' models of Newton's second law in mechanics and electromagnetism', *European Journal of Physics*, **25**, 81–89

Johansen, E. and Walter, L. (2007) 'Lean construction: Prospects for the German construction industry'. *Lean Construction Journal'*, **3** (1), 19-32

Kaplan, S. and Norton, P. (1992) 'The Balanced Scorecard – Measures that drive performance', *Harvard Business Review,* **70** (1), 47-54

Karanseh, A. and Al-Dahir, A. (2012) Impact of IT- Balanced Scorecard on financial performance: An empirical study on Jordanian banks', *European Journal of Economics, Finance and Administrative Sciences,* Issue 46, 54-70. Available at: *http://www.eurojournals.com/EJEFAS_46_05.pdf*

Keiser, PE. JD. (2012) 'Leadership and cultural change: Necessary components of a lean transformation', *Proceedings 20th Annual Conference of the International Group for Lean Construction,* San Diego, USA, 18-20 July 2012

Koskela, L. (1992) 'Application of the new production philosophy to construction'. [Technical Report No. 72] CIFE, Stanford University

Lam, E., Chan, A. and Chan, D. (2007) 'Benchmarking the performance of design-build projects: Development of project success index', *Benchmarking: An International Journal*, **14** (5), 624-635

Lancaster, G., Dodd, S. and Williamson, P. (2004) 'Design and analysis of pilot studies: recommendations for good practice'. *Journal of Evaluation in Clinical Practice*, **10** (2), 307-312.

Lantelme, E. M. V. and Formoso, C. T. (2000) 'Improving performance through measurement: The application of lean production and organisational learning principles', *Proceedings of 8th International Conference of the International Group for Lean Construction*, University of Sussex, Brighton

Leong, M. S. and Tilley, P. (2008) 'A Lean strategy to performance measurement - Reducing waste by measuring 'next' customer needs', *Proceedings of 16th International Conference of the International Group for Lean Construction*, Manchester, UK

Mitropoulos, P. and Howell, G. (2001) 'Performance improvement programs and lean construction', *Proceedings of IGLC-9*, Singapore

Moon, H., Yu, J. and Kim, C. (2007) 'Performance indicators based on TFV theory', *Proceedings of IGLC-15,* Michigan, USA

Mossman, A. (2009a) 'Why isn't the UK construction industry going lean With Gusto?', *Lean Construction Journal*, **5** (1), 24-36.

Mossman, A. (2009b) 'Creating value: a sufficient way to eliminate waste in lean design and lean production', *Lean Construction Journal*, **5**, 13 – 23.

Neely, A., Mills, J., Platts, K., Gregory, M. and Richards, H. (1996) 'Performance measurement system design: Should process based approaches be adopted?', *Int. J. Production Economics*, **46-47**, 423-431.

Nudurupati, S., Arshad, T. and Turner, T. (2007) 'Performance measurement in the construction industry: An action case investigating manufacturing methodologies', *Computers in Industry*, **58**, 667-676.

Olatunji, J. (2008) 'Lean-in-Nigerian construction: state, barriers, strategies and "go to-gemba" approach', *Proceedings 16th Annual Conference of the International Group for Lean Construction*. Manchester, UK

Rooke, J., Seymour, D. and Fellows, R. (2003) 'The claims culture; A Taxonomy of attitudes in the industry', *Construction Management and Economics*, **21** (2), 167-174

Rooke, J., Seymour, D. and Fellows, R. (2004) 'Planning for claims: An ethnography of industry culture', *Construction Management and Economics*, **22** (6), 655-662

Rooke, J. A., Koskela, L. and Seymour, D. (2007) 'Producing things or production flows? Ontological assumptions in the thinking of managers and professionals in construction', *Construction Management and Economics*, **25** (10), 1077-1085

Sarhan, S. and Fox, A. (2012) 'Trends and challenges to the development of a lean culture among UK construction organisations', *Proceedings for the 20th Annual Conference of the IGLC*, 1151-1160, San Diego, USA, 18-20 July 2012; Available at: http://iglc.net/?page_id=277

Senaratne, S. and Wijesiri, D. (2008) 'Lean Construction as a Strategic Option: Testing its Suitability and Acceptability in Sri Lanka', *Lean Construction Journal*, **5** (1), 34-48

Sikka, S., Dawood, N., Marasini, R. and Dean, J. (2006) 'Identification and development of key performance indicators to establish the value of 4D planning', *ARCOM Doctoral Workshop on Emerging Technologies in Construction*, School of the Built Environment, University of Salford: 10 November, 2006.

Suwignjo, P., Bititci, U. S. and Carrie, A. S. (2000) 'Quantitative models for performance measurement system'. *Int. J. Production Economics*, **64,** 231-241

Swain, B. and Mossman, A. (2003) 'Smooth and Lean - Hathaway Roofing Case', *Quality World*, **29** (3), 27-30

Takim, R. and Akintoye, A. (2002) 'Performance indicators for successful construction project performance', *18th Annual ARCOM Conference*, University of Northumbria: 2-4 September 2002, 545-555

Terry, A. and Smith, S. (2011) *Build Lean: Transforming construction using lean thinking.* London: CIRIA

The KPI Working Group (2000) 'KPI Report for the Minister for Construction'. [Online], Available at: www.bis.gov.uk/files/file16441.pdf (Accessed: 15 July 2011)

Wu, P. and Low S. P., (2011) 'Lean production, value chain and sustainability in precast concrete factory – a case study in Singapore', *Lean Construction Journal*, 92-109

Safety Awareness Educational Topics for the Construction of Power Transmission Systems with Smart Grid Technologies

Brian Hubbard, Qian Huang, Patrick Caskey and Yang Wang, (Purdue University, USA)

Abstract

Power transmission facilities in the U.S. are undergoing a transformation due to the increased use of distributed generation sources such as wind and solar power. The current power grid system is also antiquated and in need of substantial retrofits to make it more efficient and reliable. The new energy transmission system being designed and built to optimize power delivery is known as "Smart Grid". The increased activity in the construction of power transmission facilities and installation of new technologies into the current power system raises potential safety concerns. Existing construction management curriculum may include general information about safety training, but does not typically include information about this specialized sector. The objective of this study was to work with industry to identify hazards and safety topics appropriate for inclusion in an introductory industrial construction course. These topics were subsequently developed and incorporated into a joint undergraduate and graduate course in industrial construction. A survey of the students was performed to determine the effectiveness of the course and to obtain feedback. This paper documents information on electrical system hazards and the results of the student surveys, both of which may be considered a first step in the exploration and development of a safety curriculum to meet the needs of the future.

Keywords: Electrical Safety, Electrocution, Smart Grid, Power Transmission, Safety Education

Introduction

Construction of high voltage electrical transmission and distribution systems in the United States is rapidly increasing. (EEI 2011) Workers constructing these electrical facilities encounter hazards common to all construction sites, as well as hazards unique to sites involving high voltage currents. The construction of high voltage electrical systems poses a number of hazards including electrocution and falls; these hazards may result in fatalities and injuries. In this study, voltages exceeding 10 kv (10,000 volts) were considered high voltage installations. This reflects the range of common voltages for a substation used for power distribution to large industrial and commercial users in the United States (Durocher 2010). The purpose of this study is to determine the safety issues related to the construction of electrical transmission systems and the implementation of smart grid technologies. Construction hazards identified will be utilized in the development of educational modules for safety in an industrial construction course. A survey of the class where a portion of these concepts were implemented is presented.

Background

In recent years there has been increased construction activity in the power transmission sector (EEI 2011). The growth in this sector can be attributed to a number of factors, including:

- On-going Growth in Electricity Consumption: Electricity consumption has increased in recent years and projections forecast continued increases. Current and future energy demand in the U.S. requires new distribution substations and new power lines to connect new power sources to the power grid system to serve the increasing demand. The increase in expenditures on transmission and the distribution system over the last 3 decades has been dramatic, with an increase of over 35% from the 1980's to the 2000's (EEI 2011).

- Need for Rehabilitation of Existing Power Grid: The existing power grid system has exceeded its design life and much of the system includes outdated technology which requires reconditioning and replacement (ASCE 2011). Replacement system elements include components that can meet the increased capacities as well as components that employ new technologies such as smart grid capabilities. Smart Grid is a term used to define a transmission system that optimizes the power delivery with two-way communication between the producer and end user (DOE 2011) (Pratt 2010).

- New Energy Sources: New energy sources, including renewable "green" energy producing systems, such as wind and solar systems, are being implemented across the country. (Gellinges, 2009). These distributed energy sources often require new transmission and distribution systems. These systems are needed to connect the new power sources to the existing grid. Furthermore, the increase in the construction of new power generating sites requires an increase in the power carrying capacity of existing facilities.

The growth in the construction of transmission lines and associated power systems has created a need for an evaluation of the safety procedures in this sector. New workers in this sector will be either new to the construction industry, or from other sectors of the construction industry that have significantly different construction safety procedures. The construction workforce that transitions into the power sector from other construction sectors is not accustomed to working in the high voltage construction environment and may not be properly trained for safety in this area. In either case, the workers will not be well versed in the hazards and in appropriate hazard mitigation and safety protocol. Another consideration is that because new technologies are being implemented into these high voltage systems, there may be limited training materials available for workers and their employers.

The total number of fatalities in the power transmission and distribution industry in the U.S. is shown in Table 1 (BLS 2010). Based on the fatality statistics, approximately half of the fatalities over the last eight years can be attributed to "exposure to harmful substances or environments" which includes contact with electricity. The number of fatalities due to falls within the power transmission sector was very small over the last eight years, with many years without a fatality. In contrast, falls are the leading cause of fatalities for workers in the entire construction sector (OSHA 2012b). While this data represents those workers directly involved with power transmission and construction it does not account for the large number of deaths attributed to electrocution from workers outside the power transmission and distribution industry. For example, in 2010, 76 workers were killed due to contact with overhead power lines (BLS 2010). A better understanding of working around power transmission and distribution systems may have assisted in preventing these fatalities.

Methodology
The objective of this paper is to identify some of the safety needs of the power industry based on industry interviews and a review of regulatory requirements, identify how these needs can be translated into a curriculum for training, and evaluate the resulting safety training.

Year	Total Fatalities	Transportation Incidents	Falls	Exposure to Harmful Substances or Environments	Non-Categorized
2010	14	4 (29%)		7 (50%)	3
2009	5			3 (60%)	2
2008	21	4 (19%)		13 (62%)	4
2007	15			7 (47%)	8
2006	19	6 (32%)	4 (21%)	6 (32%)	3
2005	9	3 (33%)		4 (44%)	2
2004	25	6 (24%)	4 (16%)	11 (44%)	4
2003	14	5 (36%)		5 (36%)	4
Total	122	28 (23%)	8 (7%)	56 (46%)	30

Table1 United States Fatal Occupational Injuries for the Electric Power Transmission, Control, and Distribution Industry (BLS 2010)

Interviews with Industry

Meetings were held with three industry representatives to identify key areas where safety training was needed for electrical transmission and distribution. The interviews were intended not to provide a statistically valid sample of a large number of respondents, but rather to thoroughly investigate the issue through in-depth interviews. This approach was desirable because there are a limited number of large companies performing this work, which would make a statistically valid sample difficult to obtain. The three industry representatives consisted of an electrical contractor that specializes only in high voltage transmission line construction, an electrical contractor that provides a range of services to the construction industry including construction of sub stations for high voltage distribution (safety manager and two electricians), and a facilities owner/manager that utilizes high voltage power systems.

The main emphasis of the meetings were to determine 1) specific safety procedures for working with high voltage systems, 2) concerns regarding safety issues that need to be addressed, and 3) identify how new technologies such as smart grid systems may require additional safety procedures. While the results of these interviews may not translate to all contractors, they do provide a perspective, albeit limited, on the issues. The facilities owner provided a client perspective and also highlighted safety considerations associated with operation.

Safety Information from Regulatory Agencies

In addition to meeting with industry about construction hazards, a literature review was performed to determine the availability of regulatory requirements and safety information in the area of high voltage electrical construction. The literature review encompassed regulatory requirements in the United States, such as Occupational Safety and Health Administration (OSHA) standards, as well as availability of safety training material from other industry groups.

Identification of Construction Hazards

Upon completion of the industry interviews and review of regulatory safety requirements, a list of critical safety hazards was developed that were unique to the construction of high voltage systems. The critical safety hazards will be the focus of future training for construction personnel who may be exposed to high voltage around electrical power transmission, distribution, and control systems.

Results

The results include the findings of the interviews and the review of regulatory requirements, both of which were used to identify appropriate safety awareness educational topics.

Interviews

Interviews provided substantial information regarding hazards and the safety procedures. Key hazards and safety concerns provide a glimpse into the power industry. Interviews were conducted with a facilities manager/owner overseeing electrical improvements on high voltage building systems, a safety manager and electricians who work on electrical substations, and a constructor of high voltage power transmission systems. Discussion includes example hazards for a specific application, as well as general construction topics that can affect safety, such as the contracting method, and an example best practice to improve safety.

Hazards of Retrofitting. Retrofitting presents hazards for high voltage buildings, substations and transmission lines. For high voltage buildings, retrofitting electrical systems provides unique hazards, due to both the dangers associated with the historic system, a potential lack of documentation regarding the historic system, and an environment that may be physically constrained in terms of workspace and space for the new equipment. As an example of the safety procedure used for retrofitting an electrical system to incorporate new technologies, the procedure to isolate a 12.5kv electrical feeder system in order to install updated transformers and switchgear with advanced communication technologies was reviewed. The work procedure had both construction details and safety practices noted. The procedure required multiple managerial and safety supervisor approvals before work could be started. These approvals were required to ensure that the procedure provided adequate safety for the worker.

A discussion of the work procedure highlighted the dangers of working with old high voltage systems; old systems lack safety devices that are common on new equipment. This is illustrated by the electrical vault for power distribution that workers would have to enter, as shown in Figure 1. When working in this facility, there is a high risk of electrocution by contact due to the exposed electrical contacts. The older systems also pose a threat when activating/deactivating switches. This operation has to be completed while the system is electrified or "live". This live line work requires an electrical construction technique known as "hot sticking" to be utilized. In this method, work is done utilizing insulated tools. The worker is separated from the energized circuit by the insulated tool (Chan 2004). Even if properly insulated, the worker always has to be aware of the potential for arc flash when working on live high voltage systems. The worker may be required to wear additional personal protective equipment (PPE) such as an arc flash suit or use an arc suppression blanket (electrical blast blanket). An example of an arc flash suit with insulated electrical gloves is shown in Figure 2. This kind of suit is important in the power sector of the construction industry, but is not commonly used in other construction sectors.

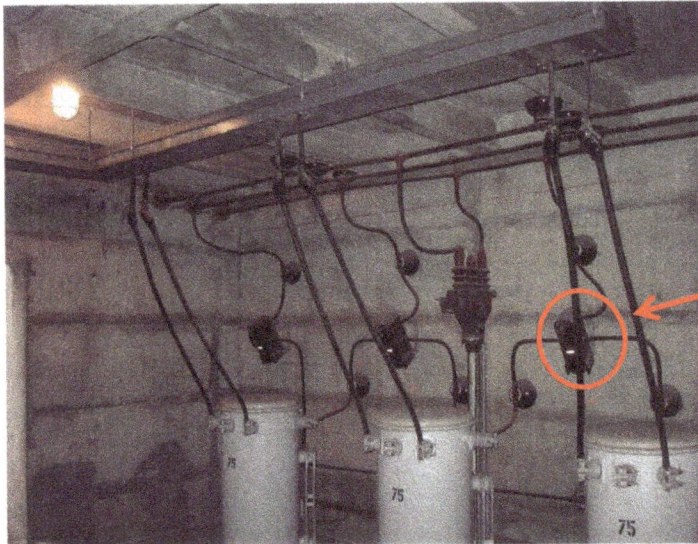

Exposed electrical contacts are an electrocution hazard near switch.

Figure 1 Outdated Electrical Vault with Switchgear and Transformers

Suit provides full body and face protection. This personal protection equipment is rarely seen in other construction sectors.

Figure 2 Example of Arc Flash Suit with Insulated Electrical Gloves

Electrical Hazards for All Workers. A major concern is not just the safety of electricians, but also the potential hazard of high voltage electrical construction for workers from other trades working near high voltage equipment. Workers in the electrical trades not only have the training, but also typically have the appropriate personal protection equipment (PPE), such as that shown in Figure 2; however, workers trained in other trades, such as carpenters, plumbers, labourers, may not understand the risks, or be wearing the PPE necessary when working in close proximity to high voltage lines. This concern points to the need to ensure that workers in all trades have a comprehensive understanding of the risks associated with working around high voltage systems.

Figure 3 shows the electrical vault after improvements have been made to the system. The advanced electrical equipment is much more protected and allows for worker access into the room without them being at risk. The new system is an excellent example of how electrical system hazards can be mitigated through the design of new systems. This system presents minimal risk to nearby workers.

Advanced technology switchgear and transformers allow system to be controlled remotely for increased safety.

Figure 3 Electrical Vault with Advanced Technology Switchgear and Transformers

Substation Hazards. Electrical substations are used to either step down the voltage of a transmission system so it can be distributed to customers or step-up voltage from power generation systems so that it can be efficiently transported on the transmission power lines. There has been increased construction of substations that step-up the voltage in order to provide transmission facilities for new energy sources. These new energy sources, such as wind and solar power, require that the voltage be stepped up to accommodate the high voltage transmission requirements. There is also significant reconstruction work being done on replace existing equipment in substations, such as the installation of new smart grid technologies, and upgrades to increase capacity to accommodate green energy sources.

High Voltage Transmission Line Construction. These systems consist of erecting the tower for the conductors, stringing the conductor wires, and connecting the conductor wires at power sources and distribution substations for step down voltage conversion. Additional new technologies to optimize the power efficiency of the grid that are being added and include: fiber optic cables for communication, wireless high voltage switches, and meteorological equipment for sensing oncoming weather issues (Brandt 2011). The process of constructing these transmission systems is a specialized field of construction and the market is dominated by a few firms. The equipment used for the stringing operation is typically custom built for the contractor based on their specifications, and is considered proprietary. The main safety concerns for the contractor focus on power lines that are not in service; although these power lines do not carry a live current, they can become energized by existing systems through either induced voltage or inadvertent contact with a live line.

In the U.S., it is difficult to obtain right-of-ways for new power transmission lines. When transmission lines are improved and the voltage of the system is increased, the power transmission authority (typically the Regional Transmission Organization, RTO) often expands the right-of-way of an existing system, which is typically more cost effective than buying new property for the lines. As a result, new systems are often built on or near existing lines, which presents hazards. Many of these existing transmission systems cannot be de-energized during the rebuilding process because the transmission system is at capacity with very little redundancy (DOE 2011). Power transmission line contractors will therefore build over the top of and around an existing energized system.

Voltage Induction Hazard. One hazard associated with electrical work is voltage induction. This is a hazard for many circumstances. For substations, hazards are often amplified because the substations are constantly being made smaller to reduce the footprint. For existing facilities, additional equipment is being added within the existing footprint, increasing the density of live electrical components. Both of these situations reduce the distances between high voltage equipment. The distance between components is an important consideration, because each piece of energized electrical equipment has a significant electromagnetic field around it that may induce voltage into other systems; this is referred to as voltage induction. In high voltage applications, the voltage induction may be high enough to cause injuries and even fatalities. To reduce these risks, special care must be taken to ground equipment and provide safe working distances (OSHA 2012).

Voltage induction is also a hazard for high voltage transmission line construction. When stringing the new power lines, the conductor may pick up voltage from the neighbouring existing lines either through induction or inadvertent contact between the de-energized line and an energized power line. Both of these hazards are difficult to prevent when working with transmission lines carrying 345kv to 750kv. The proprietary conductor stringing equipment used by these contractors all have significant capabilities to ground the wire they are stringing; this helps reduce the likelihood that the conductor will be energized during the stringing process.

Safety Challenge: Design Build Contracting. The construction contracting process can impact the safety of a construction project, in this case, substation construction. In Design Build contracting, the design is often done in tandem with the construction process. This makes it more difficult to formulate a strong construction site safety plan, since the design is evolving. On-going safety planning and effective communication is necessary to counter this.

Best practice: Job Hazard Analysis (JHA) Updates. One way to increase safety on any project is through the use of current JHA updates. One company has a policy to review the hazards of the construction site in the morning prior to the commencement of the day's work. During that time, a written job hazard analysis is provided to the worker. It should be acknowledged, however, that changes on the construction site may affect the JHA. This is especially true for construction projects that have a tight schedule due to an electrical outage. In these situations, the job hazards may be very dynamic, and change throughout the day. In this case, the JHA must be constantly modified to reflect the changing circumstances and hazards that result from a dynamic construction site. The timely updating of JHAs and effectively communicating changes to the JHA is a critical feature of a safety program.

Safety Information from Regulatory Agencies

The Occupational and Safety Health Administration (OSHA) provides the primary regulation for worker safety in the United States. While some states have their own state OSHA, these agencies typically mirror many of the requirements of OSHA. OSHA compiles data regarding occupational injuries and fatalities, by sector and by cause, and provides information to mitigate worker risk. Electrocution is one of four construction hazards that OSHA has identified as a target area in which to focus specific training, due to the number of fatalities and injuries (OSHA Training Institute 2011). OSHA has developed material for courses for both the standard OSHA 10 hour course and OSHA 30 hour course; however, this material is focused on general electrical procedures and does not provide significant details for high voltage construction.

Although not included as a standard OSHA module, there is information available. The "*OSHA Safety and Health Regulations for Construction*" provides information for working on power transmission and distribution; this information is comprehensive and covers many of the hazards discussed (OSHA 2012). As an example, a minimum clear distance to perform a "hot stick" operation is provided for power lines from 2.1kv to 765kv. The dangers of hot stick operations were mentioned in interview 1.

Another resource for safely working in high voltage transmission systems is "Live Work Guide for Substations", a resource developed by Electric Power Research Institute (EPRI) (Chan 2004). This guide is a comprehensive training manual for working on live high voltage power systems. It provides details on developing a work plan for working on live power systems, including development of JHAs, different methods for live line work, minimum approach distances for both qualified and non-qualifies workers, identification of appropriate PPE, and voltage induction hazard identification.

For the implementation of smart grid technologies, there is little safety training information available since this is a new area of the construction field. Furthermore, not all documents on the smart grid provide comprehensive information for high voltage applications. For example, a smart grid technician training guide recently written provided electrical safety hazard and PPE information, however, it did not provide extensive details on working around high voltage systems (Freeman 2010).

Safety Awareness Educational Topics

Based on the results of the interviews and the safety information from regulatory agencies, a series of topics was developed for a basic course in industrial construction. These topics are suitable for presentation to students to raise their awareness of the hazards of working around high voltage systems. Major topic areas include: 1) unique aspects of working around high voltage systems, 2) live power line work, 3) voltage induction hazards, and 4) managing a changing safety environment in a compressed construction schedule.

1. High Voltage Systems: Course material was developed to provide an introduction of how high voltage systems operate and concepts behind smart grid technologies. The step-by-step procedure to construct a high voltage transmission system is included in the course instruction. The construction operations include constructing the towers and stringing the conductors. Safety procedures are provided for each stage of the operation.

2. Live Power Line Work: A majority of the power grid construction is done either on live lines or in close proximity to live lines, which was reflected in the course material. Procedures for working on a live line are provided, along with details of the required specialized PPE and construction equipment. A review of OSHA regulations for working on high voltage systems is also included.

3. Voltage Induction: Understanding concepts related to how a de-energized line can be energized by induced voltage from electromagnetic fields from energized lines and equipment is critical for working safely in a high voltage environment. The procedures to protect workers and equipment from this hazard are included in the course instruction.

4. Managing a Changing Safety Environment in a Compressed Construction Schedule: The construction of electrical systems is typically conducted during short power outages. Some of these outages may be planned in advance and others may be the result of an emergency shutdown. In either case, the construction process for these systems is typically fast paced and very schedule driven. Working in this environment requires constant attention to safety. On-going reviews of the current job hazards must be conducted throughout the day as the job site and conditions change. In some cases, design changes may occur during a project, particularly during design build; these design changes may warrant modification of the JHA.

Course Content Development and Survey

A course was developed to serve as an introduction to the cutting-edge smart grid technologies. This course covers several interdependent areas, including power generation, energy transmission, distribution and storage, and economic policies associated with the smart grid power systems. All of these areas also addressed specific power distribution worker safety.

The first course module is an overview of the current power transmission system and basic smart grid technologies. The drawbacks of existing U.S. power grid systems are discussed along with concepts such as reliability and vulnerability, lack of end-user energy management, and two-way communication issues across the power grid are addressed. Next, the concept and definition of smart grid are elaborated, including key factors and characteristics of the smart grid. This course module is expected to deliver a comprehensive comparison between traditional power grid systems and smart grid power systems.

The second course module focuses on energy generation, distribution and storage. Since one of the important characteristics in smart grid systems is distributed energy generation, which involves multiple heterogeneous energy generation approaches (such as solar cells, wind turbines, micro-turbine generators, thermoelectric generator and fuel cells), lower the energy cost and enhance the grid reliability. Next, a review of several potential renewable energy sources is provided, including solar, wind and hydro energy generation. The module is finally concluded with a discussion on emerging technologies of energy storage with their technical challenges and trend.

The content of power grid basics and transmission line stringing methods are covered in the third module. Grid line monitoring techniques are discussed with a focus on reliability and security concerns. Superconductor based transmission line design is also introduced, because superconductors are an attractive and feasible material for the transmission line of the future. A major component of this module is the safety technique of live power line work and the associated techniques. Voltage induction was also covered under live line working because this is a common safety issue when dealing with transmission line stringing.

In order to ensure a successful implementation of smart grid technologies, the strategies and policies of U.S. power transmission systems need to be understood. Therefore, economic assessment and policies are introduced in the fourth educational module. The main reference is a report from the executive office of the President of the United States (Chopra et al. 2011). This report provides an in depth look at national policies for future Smart Grid technologies in U.S. Economic factors and considerations for implementation of smart grid systems are also discussed in this module.

Survey Data Analysis

The survey included 21 questions focused on several key smart grid concepts including: smart grid course information, benefits of smart grid technology, possible challenges of smart grid technology, and personal interest in smart grid technology.

The first six questions addressed the smart grid course information. In response to the statement, "When signing up for this class was your interest in high voltage transmission and smart grid concepts what attracted you to this class", 33% agreed. Fifty percent agreed with the statement, "After graduation are you planning on working with a firm that is actively involved with high voltage transmission/smart grid projects". In addition, 33 percent thought that they are highly likely to work on high voltage transmission/smart grid projects.

Two background knowledge related questions were assigned. Seventy-five percent agreed with the statement, "You had little knowledge of smart grid, except for some basic idea and/or definitions". The remaining 25% of respondents had no knowledge of smart grid concepts before attending this class. Of all the respondents, the ways they obtained the information on smart grid technologies were from media reports, topics from other general courses, and working experience from related projects. No participants had prior knowledge from a specific course or training module on smart grid technology. These results provided evidence to justify the need for developing a specific course that focuses on the construction of high voltage electric transmission systems and implementation of smart grid systems.

Figure 4 illustrates the topics identified as most helpful. The majority of students who completed this survey reported that distributed generation, power grid basics, and grid monitoring are most important over other topics. In addition, no respondents believe that superconductors and distributed energy resource are important.

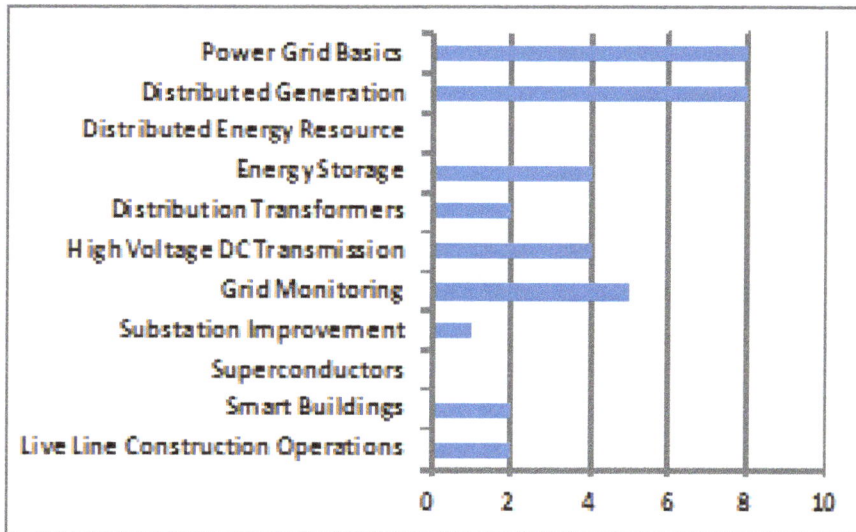

Figure 4 Survey results, list three important topics you learned from this course

The respondents also identified topics that they did not think would be used, as illustrated in Figure 5. Superconductors, distribution transformers, smart buildings and live line construction operations are not expected to be used. All respondents agree that power grid basics will be significantly used in industrial workplace and is is important for their education.

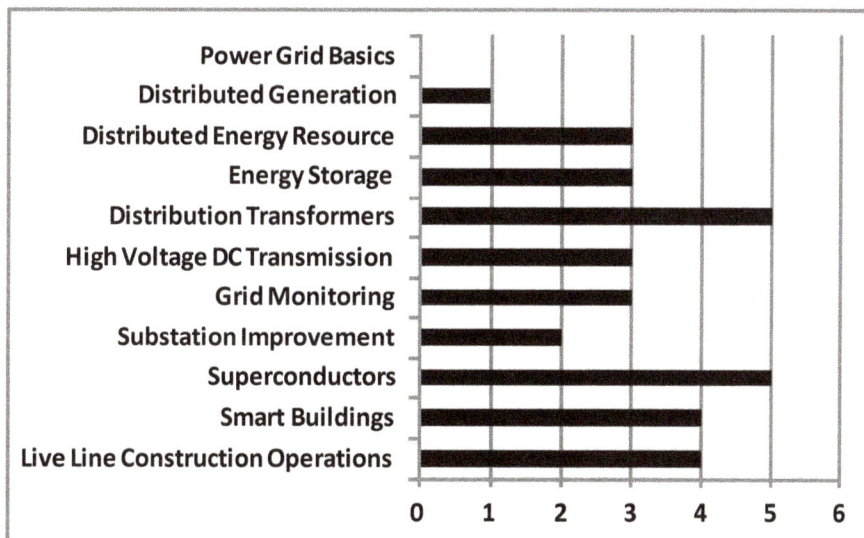

Figure 5 Survey results, list three topics you believe you will not use

Of the 21 total survey questions, 15 questions used the five-point Likert-type scale (Burt 2003). The response choices of these 15 questions are "strongly disagree", "disagree", "not applicable", "agree", and "strongly agree". Scores from 1 to 5 are assigned to responses from "strongly disagree" to "strongly agree". The 15 questions and mean response for each question are shown in Table 2. The mean response of all these questions is 3.85, which implies the average response is "agree". The mean response for questions 7 and 8 are less than 3, which indicates the participants do not have interest to do research in smart grid area. Referring to the question 15, all respondents agree that they are more likely to accept smart grid technology based on what they have learned in this class.

Survey Questions	Mean Response
(1) I am interested in learning about smart grid technology	4.33
(2) I am interested in learning about the societal advantages of smart grid technology	4.42
(3) I have access to adequate amount information about smart grid technology	4
(4) I am interested in learning about smart grid technology privacy policies	4.08
(5) I am interested in learing about smart grid technology security policies	4.08
(6) I am interested in pursuing further studies in an area related to smart grid technology	3.58
(7) I am interested in conducting research on smart grid technology	2.58
(8) I plan to become a professional in smart grid technology or related fields	2.67
(9) Knowledge of smart grid technology would benefit my future job opportunities	3.17
(10) If given the opportunity, I would use smart grid technology	4.25
(11) I would use smart grid technology to save money	4.17
(12) I would still use smart grid technology in my house even if I saw on reduction in my utility bills	3.42
(13) I am willing to use smart grid technology to help me better manage energy usage	4.25
(14) I think smart grid courses will help lead to a job	3.75
(15) Based on what you have learned in this class, are you more likely to accept smart grid technology?	5

Table 2 Survey Questions and Mean Response

Statistical methods were also used to analyse the survey data, specifically the block design approach. In this approach, different features are the interested factor, while the student factor is a nuisance factor that is measurable and controllable but not of interest. The hypothesis was that at least one of the features attracts more interest of the student. Then, the null hypothesis is that the students place equal interest on each of the fifteen features of smart grid technology. To test the hypothesis, a two-way ANOVA (Analysis of Variance) was used in this block design survey (Burt 2003). There are several assumptions put forward for this model, which include: (1) there is no interaction between treatment effect and block effect (2) errors are independent and normally distributed and variance should be constant. These assumptions will be proved and verified using the survey data. The statistical software named Statistical Analysis System (SAS) was used to performance the diagnostics and verification.

ANOVA compares the means of different question responses by using F-test approach. If the null hypothesis in this study was rejected, it indicates the difference of means among 15 questions is significant, and hence some of the topics gain more interest from the survey participants. According to statistics theory, p-value in F-test refers to the probability that the test statistic is larger than or equal to the observed statistic when the null hypothesis is true. The SAS results of the two-way ANOVA are shown in Figure 6. In this figure, since p-value (less than 0.0001 in Figure 3) is smaller than a significant small level (usually 0.05 used), null hypothesis should be rejected and our hypothesis is valid. The truth behind the survey data is some topics gain more interest from the survey participants.

Analysis of student's interest in smart grid technology
15:33 Sunday, May 27, 2012

The GLM Procedure
Dependent Variable: response

Source	DF	Sum of Squares	Mean Square	F Value	Pr > F
Model	25	107.1166667	4.2846667	5.51	<.0001
Error	154	119.8333333	0.7781385		
Corrected Total	179	226.9500000			

p-value

R-Square	Coeff Var	Root MSE	response Mean
0.471984	22.91225	0.882122	3.850000

Figure 6 Results of ANOVA procedure to test the hypothesis

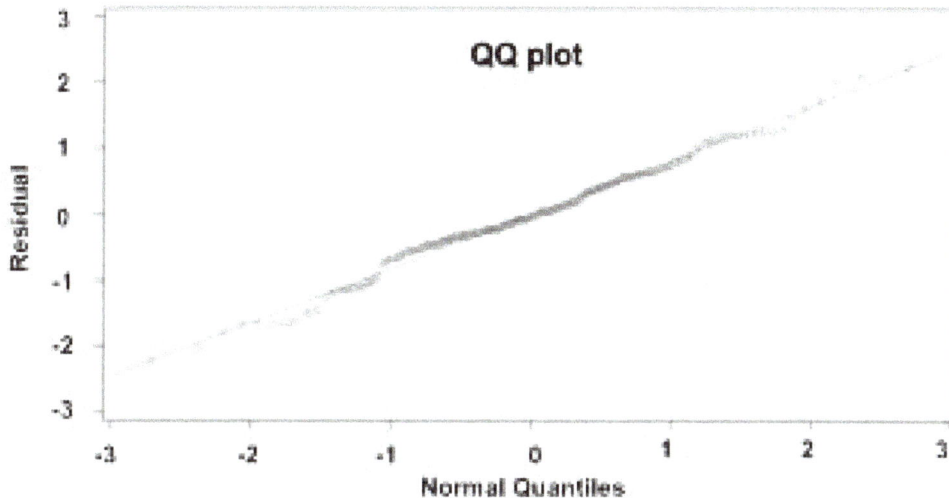

Figure 7 QQ plot for verification of assumption about normality and consistency

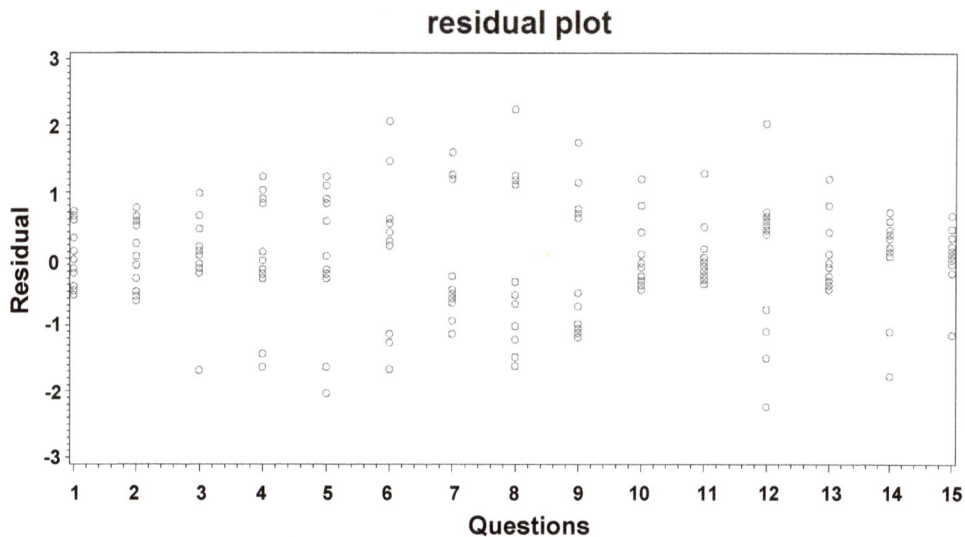

Figure 8 Residual plots for verification of independent observations

The last statistical analysis performed was to determine the topics that were most interesting to the students among the fifteen survey questions. A box plot was employed in this study. From the box plot (Figure 9), the response scores for questions 1, 2, 4, 5, 10 and 13 are more than 4. These six questions are closely related with the interest of students on learning knowledge about smart grid technology and using smart grid technology. The response scores for questions 6, 7, 8 and 9 are less than 4. These four questions are more related with the interest of students on being a professional or researcher in smart grid area. It is easy to observe from Figure 9 that the students have more intention to focus on industrial smart grid concepts rather than focusing on research in smart grid areas.

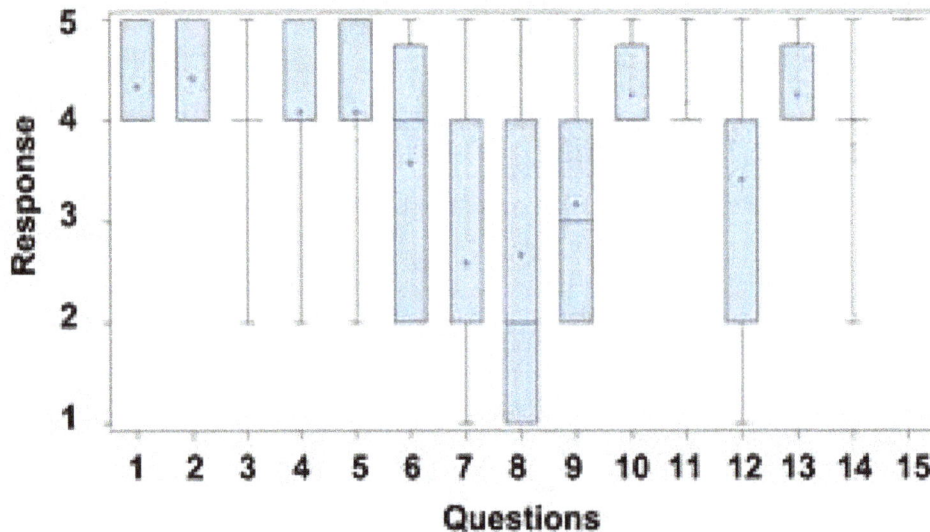

Figure 9 Analysis of student's interest in smart grid technology

Conclusion

In the United States, construction in power transmission and distribution systems is expected to be a growing sector. The many workers involved in this enormous undertaking will need to be well versed in the safety hazards involved in high voltage construction operations. This paper provides an initial identification of the kind of safety topics that new employees in this sector need to be aware of, identifies topics that may be appropriate for inclusion in an introductory industrial construction course covering construction safety of power transmission, distribution and control, and provides an evaluation of these topics from the student perspective. Additional research to more thoroughly document the safety issues and develop detailed safety curriculum to address these issues is recommended. Another area that needs to be more fully developed is the responsibility and training for supervisory personnel when workers are working on or near live electrical systems. Historically, the construction industry has proven itself very responsive to the market forces. The industry must also work to assure that the safety protocol is equally responsive to the needs of the workers.

Acknowledgements

This research study was partially supported by a U.S. Department of Energy Crossroads Smart Grid Workforce Development Program Grant.

References

ASCE, 2011, *Failure to Act, The economic impact of current Investment Trends in Electricity Infrastructure,* ASCE, Reston, VA.

Bureau of Labor Statistics (BLS) 2010, 'Fatal occupational injuries by industry and event or exposure', United States. Bureau of Labor Statistics. http://www.bls.gov/iif/oshcfoi1.htm, April 15, 2012.

Brandt, C. 2011, 'Lineman Deploy Smart Grid Technology', *Transmission and Distribution World*, **63** (9) 64N – 64Q.

Burt, R. 2003, *The factors influencing a construction graduate in deciding upon their future employer,* Journal of Construction Education, **8** (2), 94-100.

Chan, J. 2004, *Live Work Guide for Substations*, EPRI, Palo Alto, CA.

Chopra, A., Kundra, V. & Weiser, P. 2011, 'A policy framework for the 21st century grid: enabling our secure energy future', Report, Executive Office of the President of the United States, June, 2011.

Department of Energy (DOE) 2011, 'What a Smart Grid Means to Our Nation's Future', http://energy.gov/sites/prod/files/oeprod/DocumentsandMedia/Policymakers.pdf,   April   2, 2012.

Durocher, D. 2010, 'Considerations in Unit Substation Design to Optimize Reliability and Electrical Workplace Safety', *2010 Industrial & Commercial Power Systems Technical Conference*, Tallahassee, FL.

EEI 2011, *EEI Statistical Yearbook/2010*, Edison Electric Institute, Washington, D.C., http://www.eei.org/whatwedo/DataAnalysis/IndustryData/Documents/Transmission-Investment-Expenditures.pdf April 2, 2012.

Freeman, W.G. 2010, *Smart Grid Technicians Training Guide*, Educational Technologies Group, Richland, WA.

Gellinges, C. W. 2009. *The smart grid: enabling energy efficiency and demand response,* CRC Press, 1st edition, August 21, 2009.

OSHA 2012, *Safety and Health Regulations for Construction 1926.950*, Occupational Safety and Health Administration, Washington, D.C. http://www.osha.gov/pls/oshaweb/owadisp.show_document?p_table=STANDARDS&p_id=1 0821 April 10, 2012.

OSHA 2012b, *Fall Protection*, Occupational Safety and Health Administration, Washington, D.C. http://www.osha.gov/doc/outreachtraining/htmlfiles/subpartm.html April 15, 2012.

OSHA Training Institute 2011, *Construction Focus Four: Electrocution Hazards*, OSHA Directorate of Training and Education, Washington, D.C.

Pratt, R. G. 2010, *The smart grid: an estimation of the energy and CO2 benefits,* Pacific Northwest National Laboratory, Washington, www.pnl.gov/main/publications/external/technical_reports/PNNL-19112.pdf

The Benefits of, and Barriers to, Implementation of 5D BIM for Quantity Surveying in New Zealand

3

Ryan Stanley and Derek Thurnell, (Unitec Institute of Technology, New Zealand)

Abstract

Building Information Modelling (BIM) models are relational and parametric in nature, and 5D BIM is where model objects include specification data and other properties which can be directly used for pricing construction work. There is huge potential for its use by quantity surveyors (QSs) for such tasks as quantity take-offs, estimation and cost management, in a collaborative project environment. Perceptions regarding the benefits of, and barriers to, the implementation of 5D BIM by quantity surveyors in Auckland are presented, based on structured interviews with 8 QSs. Results suggest that 5D BIM may provide advantages over traditional forms of quantity surveying in Auckland by increasing efficiency, improving visualization of construction details, and earlier risk identification. However there are perceived barriers to 5D BIM implementation within the construction industry: a lack of software compatibility; prohibitive set-up costs; a lack of protocols for coding objects within building information models; lack of an electronic standard for coding BIM software, and the lack of integrated models, which are an essential pre-requisite for full inter-operability, and hence collaborative working, in the industry. Further research is recommended, to find solutions to overcome these barriers to inter-operability between 3D and 5D BIM, in order to facilitate the cost modelling process.

Keywords: BIM, Estimating, Inter-operability, Quantity surveying

Introduction

Building information modelling [or management] (BIM) is a digital representation of a building's geometric and non-geometric data, and is used as a reliable, shared knowledge resource to make decisions on a facility throughout its lifecycle (NBIMS 2010).

BIM has been in use internationally for several years, and its use continues to grow. A survey covering BIM adoption rates across North America found that 67% of engineers, 70% of architects, and 74% of contractors used BIM (McGraw-Hill Construction 2012). In the UK, the National BIM Report found that only 39% of survey respondents in 2012 were both aware of, and actually using, BIM (National BIM Survey 2013). In Australia, a survey found that 49% of architects, and 75% of both engineers and contractors used BIM, and on average, that BIM is used on 36% (engineers) to 59% (architects) of projects (BEIIC 2010). New Zealand's only national BIM survey recently found that the proportion of BIM users increased from 34 % [2012] to 57 % [2013], with a year-on-year increase in overall BIM awareness in the construction industry, from 88 % [2012] to 98 % [2013] (Masterspec 2012; 2013).

BIM extends its use to incorporate a 4th dimension (4D), 'time' and a 5th dimension (5D), 'cost,' which specifically concerns the quantification, modification and extraction of data contained within the model in order to become the primary source of information for quantity surveying (QS) services. This research focuses on the cost dimension of BIM (5D BIM), and aims to present a "snapshot" of Auckland QSs' perceptions on the benefits of, and barriers to, the implementation of 5D BIM for quantity surveying.

5D BIM

5D BIM contains objects and assemblies in the BIM model that have a cost dimension added to them, either by incorporating cost data within the BIM model objects themselves, or which can be "live linked" to estimating software tools, which is current practice in New Zealand (Boon & Prigg 2012). Parametric modelling then, facilitates the creation of a relationship between elements, and includes the specification and properties of individual elements and objects, [potentially] enabling the extraction of comprehensive and accurate information from the model which can be directly used for costing (Eastman et al. 2011). However, progress in the take up of 5D BIM is slow; in the UK, a recent survey of BIM users found that only 14% thought that BIM makes traditional bills [schedules in NZ] of quantities (BOQs) redundant (National BIM Survey 2013).

BIM and 5D in New Zealand

The use of 5D BIM in New Zealand (NZ) private quantity surveying (PQS) practices, whilst not as advanced as internationally, is developing; current use of 5D BIM for cost modelling is somewhat limited, and is restricted to certain specific aspects of cost modelling such as quantity take-offs for cost planning purposes (Stanley and Thurnell 2013). The only nationwide NZ survey found that only 8% of respondents sometimes automatically generate BOQs from their CAD models (although this may be explained, in part, by the fact that most respondents were from architectural or engineering design practices, rather than QSs). Furthermore, only 7% agreed that BIM makes traditional BOQs redundant within their organization (interestingly, far more [35%] non-BIM users agreed) (Masterspec 2013).

This limited use of 5D BIM to date is perhaps due in part to lack of use of a single BIM model in NZ; instead, projects often utilise up to three different (and separate) models, which can encompass architectural, structural and services design documentation (Boon & Prigg 2012). New Zealand's level of performance of BIM is basic, and most of the industry is still at Stage 1B ('Intelligent 3D') of the Australian Institute of Architects' (2009) BIM implementation scale. There is *some* anecdotal evidence that a *few* NZ construction projects are operating at Stage 2A: 'One-way Collaboration', where the (single) BIM model can be shared with other project participants for visualisation, coordination, communication, assessment, analysis, simulation or discipline design; however, the original model is updated in digital isolation from other discipline models (Australian Institute of Architects 2009).

Overseas studies have considered the benefits and barriers of BIM, in which estimators have been included (e.g. Sattineni & Bradford 2011; Won, Lee & Lee 2011). BIM will purportedly provide opportunities for the QS and clients by streamlining workflows and increasing the quality of cost services (Boon & Prigg 2012); however, the barriers must be understood before the potential of 5D BIM can be reached. Literature pertaining to BIM and particularly 5D BIM in NZ is limited, and so this research aims to provide a "snapshot" of Auckland QSs' perceptions on the benefits of, and barriers to, the implementation of 5D BIM.

Research Methods
Data Collection

A cross-sectional survey approach was adopted, conducted over a small time frame, which was considered appropriate, as technology tends to change quickly. The sample population were quantity surveyors, whether in private practice, or working for a contractor. Purposive, non-probabilistic sampling ensured that only those people that had some BIM experience were selected. All responses were kept confidential, and participants' anonymity was ensured. Ethics approval was sought and obtained from Unitec's Ethics Committee.

Face to face interviews gave participants the opportunity to have the wording of questions clarified, and the interviewer the ability to ensure that the questions were interpreted as intended. In order to minimise the potential for introduction of interviewer bias, the interview structure and questionnaire were piloted beforehand. The interviews were recorded which enabled post-hoc analysis of qualitative responses, in order to further reduce bias. Though a structured interview format was used, it also allowed for open ended, as well as closed questions. The format allowed the respondent to elaborate when needed, though also answer questions that were more targeted and closed, by using a semantic rating scale to assess the respondent's attitude towards the benefits of, and barriers to, implementation of 5D BIM.

Questionnaire Design

Section 1 was made up of closed questions which requested demographic information such as the number of employees in the participant's company, their role in the company, and the participant's experience with 5D BIM. No questions were mandatory; that is, there were no forced responses. Section 2 comprised 2 closed questions where respondents were asked to rate statements relating to the benefits and barriers of 5D BIM using a 5 point Likert-type semantic scale where: 1=Strongly Disagree, to 5=Strongly Agree. The items were drawn from a wide variety of sources in the literature (e.g. Popov et al. 2008; Boon 2009; Matipa, Cunningham & Naik 2010; Olatunji, Sher & Ogunsemi 2010; Samphaongoen 2010; Shen & Issa 2010; Bylund and Magnusson 2011; Sattineni and Bradford II 2011; Boon & Prigg 2012). Demographic information collected from respondents included: job role; years of QS experience; number of BIM projects worked on, and number of 5D BIM projects worked on.

Data Analysis

Participants' ratings of the benefits and barriers were analysed by identifying any general themes, if any, from the respondents' ratings. Responses to the subjective open ended questions were analysed by identifying the themes from the participant's responses and trends were identified.

Findings
Demographic Information

The demographic data collected from the eight participants is shown below in Table 1.

Participant	Job/Position	Years Experience	Number of BIM Projects	Number of 5D Projects
A	PQS*	35	0	0
B	Cntr QS** (Director)	15	1-2	0
C	Cntr QS	5	10	1-2
D	Cntr QS (Director)	30+	1-2	0
E	PQS (Director)	25	1-2	0
F	PQS	3	0	0
G	Cntr QS (Director)	18	3-4	1-2
H	PQS (Director)	24	1-2	0

Table 1 Demographic Characteristics of Sample (n=8)

*PQS: private practice QS; **Cntr QS: contractor's QS

Four participants were from private practice, and 4 from contracting firms, and had a variety of quantity surveying experience and seniority, as well as experience with projects using BIM models, wherever possible.

Benefits of 5D BIM

Participants' level of agreement (1=Strongly Disagree, to 5=Strongly Agree) with statements relating to their perceptions regarding the benefits of 5D BIM implementation for quantity surveying in New Zealand are shown in Table 2, and the findings discussed below.

Q#6.1: Visualization

Visualization was seen as beneficial to QSs; this is consistent with Samphaongoen (2010), who describes QSs as being better able to understand the project they are involved in, as they can see and interact with the 3D model. Similarly, Thurairajah and Goucher (2013) assert that the building can be viewed from any perspective in 3D, allowing QSs to make fewer assumptions about the design. As one participant stated: 'As opposed to turning over three or four hundred A2 or A1 2D drawings to try and get a picture of what the building looks like, the 3D model gives you that instantaneously.'

Q#6.2: Collaboration

5D BIM was perceived to enhance collaboration on projects, as people need to work together to make the models effective. This aligns with Popov et al. (2008) who assert that the use of 5D for cost modelling encourages collaboration on projects, and as such aids the management of the project overall. In order to achieve effective 5D, designers need to generate suitable 3D information, and this needs to be checked for clashes by the construction team. 5D software also has the ability to check for clash detection, and in this way a collaborative atmosphere is further encouraged (Won et al. 2011).

BIM depends on a collaborative approach, ideally through the use of a centralized model, where design changes are automatically updated and coordinated amongst the project team (although this is rarely achieved to date). Eastman et al (2011) assert that collaboration can be achieved by two different approaches: the first is where project teams utilize one model software from one vendor that contains all relevant design and cost information. The second approach is where project teams use proprietary or open-source software from different vendors, that contain mechanisms to ensure that data is fully exchangeable. As the software can be utilized across different disciplines, the model can be transferred between for example, QSs, architects, buildings services engineers and other consultants. This allows for real time changes to be suggested and made electronically during construction (Aranda-Mena et al 2008).

Q#	Benefits	1	2	3	4	5
6.1	The visualization of projects is increased e.g. construction details.				4	4
6.2	Collaboration on projects is enhanced as people need to work together to make the models effective.			2	6	
6.3	The quality level of the finished projects is improved as the quality of data in BIM models is maintained by its users.	1		4	2	1
6.4	Project conceptualization is made easier e.g. 3D facilitates the costing of design options during early design stage.		1	2	2	3
6.5	Increased ability to print out design details from 5D software enables greater analysis capability.	1	1	4	1	1
6.6	5D offers more efficient take-offs during the Budget Estimate Stage (i.e. $/m2 GFA).			4	2	2
6.7	5D offers more efficient generation of quantities for cost planning compared to traditional QS software and manual take off during the Detailed Cost Plan Stage (i.e. sub-elemental).			4	2	2
6.8	Earlier risk identification e.g. potential clash detection is improved, at an earlier stage than with traditional approaches.		1		2	5
6.9	Increased ability to resolve RFI's in real time.		1	2	3	2
6.10	Estimating is improved through the ability to model project options before and during construction.		1	2	2	3

Table 2 Statements Relating to the Benefits of 5D BIM for Quantity Surveying (n=8)

Q#6.3: *Project Quality and BIM Data Quality*

The quality level of the finished projects is perceived to be improved, as the quality of data in BIM models is maintained by its users. One participant (with extensive experience on projects with BIM, and also some use of 5D BIM) strongly disagreed that BIM improves data quality: *'Sometimes the quality of the data in BIM models is much reduced. A lot of objects don't have any relevant information for a QS to use.'* This highlights an issue which relates to a lack of uniformity in the way models are built, and the information they contain. The reliability of BIM estimates is dependent on the accuracy and completeness of the BIM model, which is often simplified, with minimal construction or assembly information. Standardization issues, such as when descriptions for 3D objects and the same objects in 5D software don't match is one of the reasons why QSs are not using BIM for the production and pricing of BOQs (Boon 2009).

Q#6.4: *Project Conceptualization*

Project conceptualization is perceived to be made easier with BIM, e.g. 3D facilitates the costing of design options during the early design stage. This is consistent with Popov et al (2008), who describe BIM as providing the ability to check each part of the project in relation to each of the project's options. BIM enables QS involvement in the design at an earlier stage than on traditional projects, allowing the design team to produce more design options, which enables the QS 'to quickly and efficiently produce advice to the design team and client of the cost of each option in a manner that enables direct comparison to be made' (Boon & Prigg 2012, p.7). Two participants agreed with the ability of BIM to be used at the design stage to influence the project, with one suggesting that it aids in obtaining acceptance from the client, as they are able to see the design early. Boon (2009) echoes this, saying that BIM is used at the tendering stage of projects for showing customers footage of the construction process.

Q#6.5: *Analysis Capability*

Opinion on the ability of BIM to print out design details from 5D software to enable greater analysis capability was split; this may highlight a difference of opinion between the 5D concept and the reality of 5D at present. Due to the nature of the information contained in 5D models, it is possible to use the model to print out design details, and to generate reports that are useful to other members of the project team, i.e. design changes to be made by architects, and changes to the construction program (Popov et al. 2008)., although it may be that some people are yet to determine the usefulness of BIM for data analysis in its current state.

Q#6.6: *Efficiency of Take-offs during Budget Estimate Stage*

None of the participants disagreed that BIM improves the efficiency of take-offs during Budget Estimate stage (i.e. $/m2 GFA), which suggests some perceived benefits for QSs with BIM during the Budget Estimate stage. The extraction of quantities for preliminary budget estimating is relatively simple, but it is critical that the QS identifies items missing from the model at the time of extraction (Boon and Prigg 2012). 5D BIM can provide a high level of cost detail which can be useful in the early design stages, and certain software providers are now making it possible to develop detailed cost plans by live linking the model to a 5D cost library (Thurairajah & Goucher 2013).

Q#6.7: *Efficiency of Cost Planning during Detailed Cost Plan Stage*

No participants disagreed that 5D BIM offers more efficient generation of quantities for cost planning compared to traditional QS software and manual take off during the Detailed Cost Plan Stage (i.e. sub-elemental level). One participant stated *'You can do a whole complete building harvest just by pushing a button. You can get the quantities any way you want it basically.'* Another stated *'An external wall that might have taken a good couple of hours to measure could be measured in about three mouse clicks'*. Despite this, participants also noted that extensive bulk checking is required to ensure the quantities are correct, which in turn reduces the efficiency gained: *'a lot of traditional bulk checking still needs to be done'.* Although the ability to automatically extract quantities from the BIM model reduces the time required to generate cost plans, the extraction of quantities is extremely complex due to the model containing unreliable information and an expert is often required to operate the resource (Monteriro and Martins 2013).

A quantitative study found that even when detailed estimates are produced by relatively inexperienced estimators, 5D was more effective than that of the traditional 2D estimating methods, especially with a reduction in errors and time taken (Shen and Issa 2010). This provides a further benefit, as based on these findings it is thought that firms with intermediate level staff (relatively early on in their careers) are able to be efficient at cost planning when using 5D as opposed to 2D. Moreover, where parts of a QSs role require a lot of time to process the work, BIM is able to process vast amounts of data relatively quickly and has the potential to make work easier (Samphaongoen 2010).

5D BIM costing applications are theoretically contained in the model itself, using integrated cost databases embedded within 5D models, helping to streamline the work of QSs, as rather than relying on data storage external to the application they are using for costing projects, the data is able to be applied to projects and updated as required by importing up to date information. When required, the data can be used to cost the items measured from within the one piece of software. One of the benefits of these integrated cost databases is that all relevant information is stored in one location (Samphaongoen 2010). However, instead, in practice, "live linking" models to estimating platforms is done (Thurairajah & Goucher 2013). This also seems to be the case in New Zealand at present (Boon & Prigg, 2012; Stanley & Thurnell 2013).

Q#6.8: Risk Identification
Seven of 8 participants agreed that BIM offers earlier risk identification e.g. potential clash detection is improved at an earlier stage than with traditional approaches. These findings seem consistent with a study in Korea where it was found that clash detection was used in over 70 % of projects (Won et al 2011). This indicates a clear link between BIM and the ability to reduce risk on projects.

The importance of identifying risks early on in projects is thought to be a vital element to a project's success. One participant supported this *'Clash detection is key for us, everything relates to time.'* This echoes Thurairajah and Goucher (2013), who state that clash detection is a key benefit of BIM for cost consultants. The use of BIM to reduce risk is supported by Boon (2009) where QSs are able to analyze risk earlier and derive other construction options. By finding problems early, it may be possible to save both time and money.

Q#6.9: Ability to Resolve Requests for Information (RFIs) in Real Time
Five of 8 participants agreed that BIM enables the increased ability to resolve RFI's in real time. This suggests that some projects in Auckland are able to use BIM during the construction phase, and not just during the design phase for visualization purposes. This is supported by Ghanem and Wilson (2011) who discuss a case study on a project which demonstrated that by using BIM they were able to save money by detecting clashes and therefore avoid RFIs.

Q#6.10: Estimating and Project Options
Only 1 participant disagreed with the notion that estimating is improved through the ability to model project options before and during construction. By considering project options early, fewer variations are likely to occur during construction. One participant described how when considering different project options in 5D, *'You can see the quantities change and can update based on that.'* The ability to update and change quantities quickly can be a major benefit for QSs in terms of cost modelling. Another commented: *'If an architect does a new*

model every week and you've got it linked up with a cost plan you can actually have the dynamic link so that it updates your quantities with the new model.'

Olatunji et al (2010) suggest that BIM allows professional QSs to identify factors that have economic benefit or consequence on various design options in order to select the most suitable and cost efficient proposal. Furthermore, early design advice 'should lead to increased client satisfaction as they are receiving earlier economic feedback on the alternatives available' (Thurairajah & Goucher 2013, p.3).

Barriers to 5D BIM Implementation

Participants' level of agreement (1=Strongly Disagree to 5=Strongly Agree) with statements relating to their perceptions regarding the barriers to 5D BIM implementation for quantity surveying in New Zealand are shown below in Table 3, and the findings are then discussed.

Q.#	Barriers	1	2	3	4	5
7.1	Lack of software compatibility restricts its use.			2	3	3
7.2	The setup cost inhibits its use i.e. software, training and hardware costs.			2	4	2
7.3	Increased risk exposure discourages companies e.g. legal issues such as ownership of BIM models.		2	3	2	1
7.4	Cultural resistance in companies hinders its effectiveness.		2	2	1	3
7.5	Incompatibility with industry recognized element formats for cost planning prevents companies from adopting the software (e.g. the NZIQS "Elemental Analysis of Costs of Building Projects").	1	3	1	3	
7.6	Incompatibility with current Standard Methods of Measurement (i.e. "NZS 4202:1995"), prevents firms from adopting the software for SOQ production.	1	1	3	1	2
7.7	Lack of integration in the model decreases the reliability and effectiveness of 5D (e.g. Arch./Eng./MEP designers are not all working off the same model).		1	1	3	3
7.8	Lack of protocols for coding objects within BIM models by designers hinder the development of cost modelling using BIM (e.g. lack of complete specification information in BIM models inhibits accurate quantity generation for estimating).			2	2	4
7.9	Some companies feel their current software meets their needs, so see no need to change.		1	3	3	1
7.10	The fragmented nature of the construction industry limits the potential of BIM.	1	1	1	4	1
7.11	Lack of an electronic standard for coding BIM software to Standard Methods of Measurement limits the potential of BIM for cost modelling.			2	2	4

Table 3 Statements Relating to Potential Barriers to 5D Implementation (n=8)

Q#7.1: Software Compatibility

No participants disagreed with the notion that a lack of software compatibility restricts the use of BIM, which may indicate that lack of inter-operability is a barrier to the use of 5D BIM for quantity surveying. One participant commented *'It's not about physical compatibility of the software, it's about the knowledge about working with different software.'* Another commented *'If a lead architect is using Archicad and everybody else is using Revit there's still that gap which makes it very difficult for estimating in 5D.'*

Inter-operability is the smooth exchange of information across all BIM disciplines involved which is required to maximize the benefits that BIM offers (Thurairajah & Goucher 2013).

However, fuelled by the fragmented and isolated construction industry, vendors often run software in proprietary type formats that restrict the exchange of critical building data between multiple organisations, and such incompatibility between the BIM model and estimating platforms is seen as a major barrier to 5D BIM implementation (Olatunji 2011). In an attempt to overcome this challenge, advancement is being made to improve the inter-operability of data exchange between BIM models and costing tools through open data standards such as Industry Foundation Classes (IFCs). IFC standards have been generated by the International Alliance of Interoperability (IAI) to help govern the exchange of data between CAD software tools, estimation software tools and other construction application software tools by creating a neutral file format. IFCs are believed to be important for cost consultants, as without complete inter-operability, items will be missed from the BIM model as they are combined, and therefore missed from estimates and schedules of quantities. However, there are still compatibility issues associated with IFC type files that industry is currently trying to work through (Thurairajah & Goucher 2013

Q#7.2: 5D Setup Costs, i.e. Software, Training & Hardware Costs

Six of 8 participants agreed that the setup cost of 5D inhibits its use i.e. software, training and hardware costs. Software and hardware upgrades are considered as significant barriers to BIM implementation, particularly for small-medium enterprises (SMEs) (McGraw-Hill Construction 2012). One participant alluded to big companies being able to meet the setup costs, however for *medium tier or smaller companies it's fairly expensive'*. Thurairajah and Goucher (2013), in a survey of quantity surveyors about the benefits and barriers of 5D BIM implementation, found that most of the respondents indicated a strong training requirement associated with BIM implementation, which, although time-consuming and difficult (as only a number of users have expert knowledge in the resource), is considered critical to BIM's adoption.

Q#7.3: Risk Exposure

There was only slight agreement overall that increased risk exposure discourages companies', e.g. legal issues such as ownership of BIM models, however 3 participants seemed undecided. The legal issues such as who has rights to the information contained in the BIM models, who is in charge of the information that is in the model, what happens when there are errors in the model and other responsibilities that relate to the model need to be addressed (Boon 2009). Klein (2012) concurs, and reports 'before the full potential of BIM can be released with parties working in collaboration, there needs to be an innovation in contracts and insurances that underwrites stakeholders for financial loss' (p.14).

Q#7.4: Cultural Resistance

Only 2 of 8 participants disagreed with the notion that cultural resistance in companies hinders BIMs effectiveness for cost modelling. Participants made reference to the traditional nature of the industry, one surmising *'We're basically going from horse drawn carts to motor vehicles.'*

A recent case study in New Zealand related how several BIM-capable project participants were not prepared to share BIM information between firms (Brewer, Gajendran & Runeson 2013). This type of culture or dynamic on projects may pose another barrier to successful BIM adoption and use for 5D BIM by QSs, and cultural transformation is a much greater challenge than any technological challenge arising from BIM; there is some reluctance from older QS employees to use 5D BIM, but younger employees are much more optimistic (Boon & Prigg 2012).

Q#7.5 Incompatibility with Industry Recognized Cost Planning Element Formats

Only 3 participants agreed with the notion of 5D BIM's incompatibility with the industry recognized elemental format for cost planning, which prevents companies from adopting the software (e.g. the NZIQS 'Elemental Analysis of Costs of Building Projects'). Shen and Issa

(2010) found that estimating using 3D software when contrasted with traditional 2D estimating resulted in reduced errors and time taken.

However, Boon and Prigg (2012) report that BIM models currently contain numerous design errors and often have important information missing from them, which hinders BIM's use for producing 5D cost services, as the data is too incomplete or inaccurate to use.
In order to provide cost planning services when using BIM, there is a need for BIM models to be correct, complete and objects must contain all the data needed, which at present is not the case in New Zealand. Furthermore, significant time is needed to pick up what is not shown in the models by reviewing 2D drawings that show the missing building items (Stanley & Thurnell 2013).

Q#7.6: Incompatibility with Current Standard Methods of Measurement (SMM)

There was a great deal of mixed opinion between the participants on this issue of BIM's incompatibility with current Standard Methods of Measurement (i.e. 'NZS 4202:1995'), preventing QS firms from adopting 5D software for BOQ production. One participant asserted that NZS 4202:1995 is incompatible with 5D, '*except for about 5 trades, like Blockwork, Brickwork, Concrete and maybe Suspended Ceilings. However everything else are pretty much composite items.*' This sentiment is common in the literature; Matipa et al (2010) suggest that current Standard Methods of Measurement were developed for more paper based surveying. In New Zealand, there is little use of 5D BIM to produce efficient BOQs; Stanley & Thurnell (2013) report 'few participants agreed that there is currently an increased use of 5D BIM for the production and pricing of Schedules (Bills) of Quantities (SOQs) during tender/bid stage' (p. 5). However, Boon and Prigg (2012) assert that marginal benefits can be achieved through the extraction of certain building items such as doors, windows, volumes of concrete, steelwork quantities, and services trades.

Q#7.7: Lack of Integration in the Model

Six of 8 participants agreed that a lack of integration in BIM models decreases the reliability and effectiveness of 5D (e.g. where each design discipline develops their own BIM model in isolation from each other). One participant said '*We find it's all time and cost related. Initial models we receive are poorly integrated with each other, we spend a lot of time making it integrated.*' Boon and Prigg (2012) assert that a balance needs to be found between the information architects need to use to build the 3D models, and the additional information needed for QSs to model the costs in projects. This underlying issue - lack of integration, where parties in the industry are said to work separately, and as a result this also separates the information required for BIM - is thought to be a major barrier to 5D BIM implementation (Bylund & Magnusson 2011).

Q#7.8: Lack of Protocols for Coding BIM Objects

Six of 8 participants agreed that a lack of protocols for coding objects within BIM models by designers hinders the development of 5D BIM. The need for a coding standard for 5D was highlighted by one participant '*I think a standard is probably needed, no different to the measuring standard that we have, just so that it can be standardized throughout the industry.*' The Royal Institution of Chartered Surveyors (RICS) in the UK have worked with industry to develop new rules of measurement (NRM) which will facilitate 5D BIM, and are extending this collaboration with the Australian Institute of Quantity Surveyors in Australia (buildingSMART 2012). Currently in New Zealand there are no standards that facilitate the embedment of design data to ensure extracted quantities are compliant with the quantity surveyor's SMM (e.g. NZS 4202:1995), but a technical sub-committee of the New Zealand Institute of Quantity Surveyors (NZIQS) is attempting something similar by proposing the use of the Association of Coordinated Building Information in New Zealand's (ACBINZ) Coordinated Building Information (CBI) classification system to revise New Zealand's standard method of measurement. The CBI classification system was created to coordinate information sources such as drawings, specifications, quantities, technical and research information and publications (Masterspec, 2012). The NZIQS sub-committee came to their conclusion on the basis that it was a similar coding system to the one used in Singapore, the Construction Electronic Measurement Standard (CEMS), a classification system established for BIM measurement that is globally recognised as being successful (Boon & Prigg 2012).

Q#7.9: Current Software Meets Needs

Only one of 8 participants disagreed that some companies feel their current software meets their needs, so see no need for change. This may suggest that smaller QS firms perceive 5D not to be a viable option at present. This characteristic of smaller consultant firms is shown in a study of small-medium enterprises (SMEs) in the UK Institution of Structural Engineers, which found that 73% of respondents think that BIM implementation presents serious cost and commercial challenges, and 76% of small (less than 10 employees) firms are not BIM-experienced, and so have little understanding of the finer details (Office Insight 2013).

Q#7.10: Fragmented Nature of the Construction Industry

Only 2 of 8 participants disagreed that the fragmented nature of the construction industry limits the potential of BIM. Masterspec (2012) sees this as one of the central barriers to BIM implementation, and suggests that a shift in current workflows is required. Olatunji et al. (2010) assert that BIM, and in particular 5D BIM, requires the collaboration, database integration and commitment of companies to the use of BIM software, and that as these areas are still in a separated and fragmented state, it further limits the effectiveness of 5D BIM.

Q#7.11: Lack of an Electronic Standard for Coding BIM Software

There was a high level of agreement from participants that a lack of an electronic standard for coding BIM software to Standard Methods of Measurement limits the potential of 5D BIM. One participant indicated the need for an electronic standard for BIM by saying, 'Often the designers don't code everything and if they code it they can code it incorrectly. With software, to some extent it's only as good as the information that's input in the first instance.' Although BIM-assisted estimating tools can generate large quantities of construction items in order to efficiently produce cost estimates, the extracted quantities have a lack of understanding of construction methods and procedures, which reduces the accuracy of estimates (Shen & Issa, 2010). It is these kinds of issues that a common electronic coding standard for BIM would need to address. The NZ Government's Productivity Partnership is working through the National Technical Standards Committee (NTSC) to produce an online BIM handbook for New Zealand, as part of its strategy to accelerate the application of BIM in construction here. NTSC has commissioned NATSPEC of Australia to write the New Zealand BIM handbook, currently due for release for industry comment, as well as electronic exchange standards (BIM Handbook in Production 2013).

Conclusions

The perceptions of a sample of Auckland quantity surveyors on the benefits of, and barriers to, the implementation of 5D BIM have been identified. Findings suggest that 5D-BIM may provide advantages over traditional forms of quantity surveying (in Auckland) by increasing efficiency, increasing visualization of construction details, and earlier risk identification. However there are perceived barriers to 5D-BIM implementation within the construction industry: a lack of software compatibility; prohibitive set-up costs; a lack of protocols for coding objects within building information models; lack of an electronic standard for coding BIM software, and the lack of integrated models, with objects containing full and complete data required to fulfil cost modelling tasks efficiently ,which are an essential pre-requisite for full inter-operability, and hence collaborative working, in the industry. As currently practised, 5D BIM takes place outside the core BIM model by live linking it to a third party estimating software. Participants had doubts for the feasibility of level 3 full collaborative BIM that contains integrated cost data, suggesting that the ultimate goal of BIM may never eventuate.

However, there was a strong indication that 2D drawings would eventually succumb to BIM in the future.

Some participants noted that in the future, such barriers are likely to be overcome by increasing cross-disciplinary collaboration on BIM modelling, allowing 5D BIM use to become more prominent. It is thought that as the use of BIM increases, a cultural change will take place, and 5D BIM will increasingly be more widely used by quantity surveyors in the Auckland construction industry for cost modelling. Although (due to the small sample size) these findings are not generalizable to the New Zealand quantity surveying population as a whole, they do provide a 'snapshot' of current opinion on the benefits of, and barriers to, the implementation of 5D BIM in Auckland. The accelerating implementation of BIM means that these perceptions are likely to change in the future, and this research provides a benchmark against which to gauge changes in the use of 5D BIM for cost modelling, which could help find solutions to overcome these barriers to inter-operability between 3D and 5D BIM, and report on the opinions of industry to the solutions once they have been implemented.

References

Aranda-Mena, G., Crawford, J., Chevez, C., & Froese, T. (2008) 'Building information modelling demystified: Does it make business to adopt BIM?' *CIB W78 2008 International Conference on Information Technology in Construction,* Santiago, Chile

Australian Institute of Architects (2009) 'Towards Integration. National Building Information Modelling (BIM) Guidelines and Case Studies', *Cooperative Research Centre for Construction Innovation (CRCCI)*, viewed 11 May 2013 http://www.construction-innovation. info/images—/pdfs/Brochures/Towards_Integration_Brochure_170409b.pdf

BEIIC (2010) 'Productivity In The Buildings Network: Assessing The Impacts Of Building Information Models', *Built Environment Innovation and Industry Council (BEIIC)*, Melbourne, Australia

BIM Handbook in Production (2013) *Building & Construction Productivity Partnership*, viewed 29 November 2013 http://buildingvalue.co.nz/news-events/bim-handbook-production

Boon, J. (2009) 'Preparing for the BIM revolution', *13th   Pacific Association of Quantity Surveyors Congress (PAQS 2009)*, viewed 12 April 2013 http://rismwiki._vms.my/images /7/72/PREPARING_FOR_THE_BIM_REVOLUTION.pdf

Boon, J., & Prigg, C. (2012) 'Evolution of quantity surveying practice in the use of BIM – the New Zealand experience', *Joint CIB International Symposium of W055, W065, W089, W118, TG76, TG78, TG81 and G84*, viewed 5 July 2013 http://www.irbnet.de/daten/iconda/CIB_ DC25601.pdf

Brewer, G., Gajendran, T. & Runeson, G. (2013) 'ICT & innovation: A case of integration in a regional construction firm', *Australasian Journal of Construction Economics and Building,* **13** (3), 24-36

buildingSMART Australasia (2012) 'National Building Information Modelling Initiative', *Vol.1*, viewed 4 July 2013 http://buildingsmart.org.au/nbi-folder/NationalBIMIniativeReport_6June 2012.pdf

Bylund, C. & Magnusson, A. (2011) 'Model based cost esimations–an international comparison', viewed 4 July 2013 http://www.bekon.lth.se/fileadmin/byggnadsekonomi/Carl Bylund_AMagnusson_Model_Based_Cost_Estimations_-_An_International_Comparison_ 2_.pdf

Eastman, C., Teicholz, P., Sacks, R. & Liston, K. (2011) *BIM Handbook: A Guide to Building Information Modeling for Owners, Managers, Designers, Engineers and Contractors*, John Wiley and Sons, NY

Ghanem, A. A. & Wilson, N. (2011) 'Building information modelling applied on a major csu capital project: A success story', *47th ASC Annual International Conference*, viewed 30 May 2013 http://ascpro0.ascweb.org/archives/cd/2011/paper/CPGT274002011.pdf

Klein, R. (2012) 'A work in progress', *RICS Construction Journal*, Feb-Mar 2012, 14

Masterspec. (2012) *New Zealand National BIM Survey 2012*, viewed 10 May 2013 http://www.masterspec.co.nz/news/reports-1243.htm

Masterspec. (2013) *New Zealand National BIM Survey 2013*, viewed 22 November 2013 http://www.masterspec.co.nz/news/reports-1243.htm

Matipa, W.M., Cunningham, P. and Naik, B. (2010) 'Assessing the impact of new rules of cost planning on building information model (BIM) schema pertinent to quantity surveying practice', *26th Annual ARCOM Conference*, viewed 16 October 2013 http://web.itu.edu.tr/~yamanhak/yayin/p2010b.pdf

McGraw-Hill Construction (2012) *The Business value of BIM in North America: Multi-Year Trend Analysis and User Ratings (2007-2012)*, McGraw-Hill Construction, New York

Monteiro, A. & Martins, J.P. (2013) 'A survey on modeling guidelines for quantity takeoff-oriented BIM-based design', *Automation in Construction, 35*, 238-253

National BIM Survey (2013) 'National BIM Report 2013', viewed 22 November 2013 http://www.thenbs.com/pdfs/NBS-NationlBIMReport2013-single.pdf

NBIMS (2010) 'National Building Information Modeling Standard', viewed 11 November 2013 http://www.wbdg.org/pdfs/NBIMSv1_p1.pdf

Office Insight (2013) 'Government unveils BIM initiative for SMEs as survey reveals small business concerns', viewed 29 November 2013 http://workplaceinsight.net/government-unveils-bim-initiative-for-smes-as-survey-reveals-small-business-concerns/?goback=.gde_88902_member_274944403

Olatunji, O.A. (2011) 'Modelling organizations structural adjustment to BIM adoption: A pilot study on estimating organizations', *Journal of Information Technology in Construction, 16*, 653-668

Olatunji, O.A., Sher, W., Ogunsemi, D.R., (2010) 'The impact of building information modelling on construction cost estimation', *W055 - Special Track 18th CIB World Building Congress*, May 2010, Salford, UK

Popov, V., Migilinskas, D., Juocevicius, V. and Mikalauskas, S. (2008) 'Application of building information modelling and construction process simulation   ensuring   virtual project development concept in 5D environment', *25th International Symposium on Automation and Robotics in Construction*, viewed 12 March 2013 http://www.iaarc.org/publications/fulltext/7_sec_090_Popov_et_al_Application.pdf

Samphaongoen, P. (2010) 'A visual approach to construction cost estimating', viewed 17 October 2013 http://epublications.marquette.edu/theses_open/28

Sattineni, A. & Bradford II, R. H. (2011) 'Estimating with BIM: A survey of US construction companies', *Proceedings of the 28th ISARC*, viewed 12 March 2013 http://www.iaarc.org/publications/proceedings_of_the_28th_isarc/estimating_with_bim_a_survey_of_us_construction_companies.html

Shen, Z. & Issa, R.R.A. (2010) 'Quantitative evaluation of the BIM-assisted construction detailed cost estimates', *Journal of Information Technology in Construction, 15*, 234-257, viewed 12 March 2013 http://www.itcon.org/2010/18

Stanley, R. & Thurnell, D. (2013) 'Current and anticipated future impacts of BIM on cost modelling in Auckland'. *Proceedings of 38th AUBEA International Conference*, Auckland, New Zealand, 20-22 Nov 2013

Thurairajah, N. & Goucher, D. (2013) 'Advantages and challenges of using BIM: A cost consultant's perspective**',** *49th ASC Annual International Conference*, viewed 20 November 2013 http://ascpro.ascweb.org/chair/paper/CPRT114002013.pdf

Won, J., Lee, G., and Lee, C. (2011). *Comparative analysis of BIM adoption in   Korean construction industry and other countries.* University, Seoul, Korea, viewed 22 October 2013 http://biis.yonsei.ac.kr/pdf/Comparative%20analysis%20of%20BIM%20adoption%20in%20 Korean%20construction%20industry%20and%20other%20countries.pdf

Internationalisation of Construction Business and E-commerce: Innovation, Integration and Dynamic Capabilities

4

Thayaparan Gajendran, (The University of Newcastle, Australia)

Graham Brewer, (The University of Newcastle, Australia)

Malliga Marimuthu, (Universiti Sains Malaysia, Malaysia)

Abstract

The role of internet and web based applications in delivering competitive advantage through e-business process is widely acknowledged. However, little is done by way of research to use the dynamic capability framework to explore the role of ecommerce in the construction business internationalisation. The aim of this paper is to present a literature based theoretical exploration using dynamic capability view to discuss internationalising construction businesses through electronic commerce (e-commerce) platforms. This paper contextualises the opportunities for internationalising construction, using a mix of supply chain paradigms, embedded with e-commerce platforms. The discussion concludes by identifying the potential of dynamic capabilities of a firm to exploit the innovation and integration potential of different e-business systems, in contributing to the internationalisation of construction businesses. It proposes that contracting firms with developed dynamic capabilities, has the potential to exploit e-commerce platforms to channel upstream activities to an international destination, and also offers the firm's products and services to international markets.

Keywords: Dynamic Capabilities, Innovation, E-commerce, Building Information Modelling (BIM), Integration, Internationalisation

Introduction

International business is characterised by any form of transaction taking place across national borders for the purpose of satisfying the needs and demands of individuals and firms (Rugman and Collinson 2009). The opportunities arising from globalisation, while elevating competition in domestic markets, provide construction firms with access to international markets. A number of construction firms already operate in international markets, trading their design services[i] (Reina and Tulacz, 2010a) and construction products or services[ii] (Reina and Tulacz, 2010b), amounting to significant monetary value. Internationalising a construction business is a complex process involving decisions on what international region, country or market to enter; how to make the international market entry (as exports-imports or foreign direct investments) and what is the best-fit business model(s) for gaining sustained competitive advantage (See Rugman and Collinson, 2009; Howes and Tah 2003; London, 2010).

Construction firms (e.g. contracting and consulting firms including architectural and project management firms) could exploit international markets in at least two forms: (a) outsource their selected core or non-core business functions or operations to an international operator (supplier focus) and/or (b) offer the firm's products or services in the international market (customer focus). Firms can choose to internationalise their business via an import or export mode or foreign direct investment mode (FDI) (Menipaz and Menipaz 2011). Construction firms may view their core business as being dominated by knowledge or design (e.g. architect, specialist design, project management, management contracting services etc.), manufacturing or production (e.g. building components production: lifts, escalators etc.), assembly (e.g. fabricators, principal contractors, labour sub-contracting services etc.) or a hybrid of some, or all, of the above.

Construction firms, in internationalising their business, need to be innovative and fully understand their capabilities, specifically their 'dynamic capabilities' (Teece, 2007), which enable them to sustain competitive advantage. Dynamic capability is defined as those capabilities that "operate to extend, modify or create ordinary capabilities to give competitive advantage" (Winter, 2003: 991). The ability of appropriate resources and capabilities (Teece, Pisano and Shuen, 1997; Daniel and Wilson 2003) for the skilful design and execution of an e-business model (Wu and Hisa, 2008) contextualised through the supply chain, can offer firms the desired competitive edge (Ash and Burn, 2003; Smart, 2008; Roy, Sivakumar and Wilkinson, 2004).

Lambert and Cooper (2000: 65) indicates that "one of the most significant paradigm shifts of modern business management is that individual businesses no longer compete as solely autonomous entities, but rather as supply chains". Supply chains are primarily focused on how the firm delivers its products and services to clients via effective flow of material, plant, people, finances and information. Therefore, the need to conceptualise the design and operations of a business from a supply chain perspective, has gradually gained significant attention (Min & Zhou, 2002; Cutting-Decelle et al., 2007; Vrijhoef and Koskela, 2000). The evolution of supply chain paradigms is coupled with developments in the Information and Communication Technologies (ICT) and vice versa (Pant, Sethia, and Bhandarib, 2003; Donk, 2008). Historically ICT developments in the organisations moved from the 'automation agenda' to 'inter firm integration' and then to 'supply chain wide integration' (Show, 2000; Fawcett, and Magnan, 2002; Fawcett et al., 2007) while the supply chain integration agenda focused on exploiting ICT/e-commerce developments for improved communication, customer relationship management, demand management, production management etc. (Donk, 2008). The mutual aim of both e-business and supply chain management, are about performing effective business transactions between the trading partners through sharing of business information and developing or maintaining good business relationships (Zwas,s 1996; Min & Zhou, 2002). Therefore, blending the alternative e-commerce models with supply chain paradigms (see Smart 2008) is critical in unearthing and exploiting the dynamic capabilities of a firm (Ash and Burn, 2003) and these efforts have been emphasised for global supply chain by Eyob and Tetteh (2012).

This paper explores the possible dynamic capabilities that construction firms can marshal through developing alternative e-business models, contextualised through the supply chain perspectives, to internationalise their business. In doing so, this paper also evaluates the issues that are beyond the control of the firms, impacting internationalisation using ecommerce platforms.

Dynamic Capabilities: a Synopsis

Teece (2007) suggests that firms operating in globally competitive environments with geographically dispersed operations require more than the ownership of difficult-to-replicate assets to attain sustainable advantage. He suggests that such firms 'also require unique and difficult-to-replicate *dynamic capabilities'*. Teece and Pisano (1994) term dynamic as:

> *the shifting character of the environment; certain strategic responses are required when time-to-market and timing is critical, the pace of innovation accelerating and the nature of future competition and markets difficult to determine. The term capabilities emphasises*

the key role of strategic management in appropriately adapting, integrating and reconfiguring internal and external organisational skills, resources and functional competencies toward the changing environment. (p1)

Dynamic capabilities are about a firm's ability to deploy resources or capabilities in effective combinations and modify its specific organisational processes to achieve its goals where the resources and capabilities can be tangible and intangible (see also Makadok 2001). The nature of the dynamic capabilities is well explained as an extension of resource based view (RBV) that describes the conditions under which firms, based on their bundles of resources and capabilities may achieve a sustained competitive advantage (Barreto, 2010). Eisenhardt & Martin (2000) describes dynamic capabilities as:

The firm's processes that use resources – specifically the processes to integrate, reconfigure, gain and release resources – to match and even create market change. Dynamic capabilities are therefore the organisational and strategic routines by which firms achieve new resource configurations as markets emerge, collide, split, evolve and die. (p.1107)

Ambrosini and Bowman (2009: 31) suggest Teece, Pisano and Shuen (1997) and Nelson and Winter (1982) "take an efficiency approach to firm performance rather than a privileged market position approach (the latter being the underpinning for Porter's (1980) theory of competitive advantage)". Porter's (1990) Competitive Advantage Theory proposes that any firm that understands and manages the effects of the five major factors, namely: demand conditions, presence or absence of supporting suppliers, degree of rivalry, threat of new entrants and threat of substitutes, will posses significant competitive advantage over competitors. The proponents of the dynamic capability approach places emphasis on the internal factors of the firm (rather than external factors) contributing to competitive advantage (Ambrosini and Bowman, 2009).

Teece's (2007) conceptualisation of dynamic capabilities comprises three distinct processes (or routines), namely *sensing, seizing and reconfiguring the resource base*. Sensing opportunities (and threats) is about scanning the environment (e.g. markets, technological advancements etc) to identify new opportunities (Teece, 2007). 'Sensing' requires construction firms to maintain good relationships with trading partners, (for example, suppliers, contractors etc) and to spot related advancements that can create new opportunities. 'Seizing' opportunities is about capturing existing and emerging opportunities, and possible investments in relevant technologies (O'Reilly III and Tushman, 2008; Teece, 2007). 'Reconfiguring' the resource base is about a firm's ability to recombine its internal and external resources and operating capabilities (Teece, 2007) to create sustained competitive advantage.

The dynamic capabilities framework provides a sensible approach to analyse the e-commerce initiatives in internationalising the construction business. Daniel and Wilson (2003) argue that dynamic capabilities are critical for businesses operating through e-business models to provide them with sustained competitive advantage. They argue that two groups of capabilities are essential for e-business adoption: the first group is associated with the sensing and seizing (routines) to identify innovative approaches to design of e-business environments, while the second group relates to reconfiguring and integrating resources associated with e-business initiatives within the existing operations of the business. It is argued that both groups of capabilities can best be analysed via the supply chain context as it provides a holistic perspective to connectivity between trading partners in construction projects. Daniel and Wilson (2003) identified eight innovation and integrative capabilities mostly encapsulating the three dynamic capacity routines. Below, the dynamic capabilities (sensing, seizing and re configuring) proposed by Teece (2007) are explored through innovative and integrative capabilities proposed by Daniel and Wilson (2003) in the context of e-business transformation (see also Wu and Hisa 2008; Rindova and Kotha, 2001).

1. Innovation in culture and climate: ability of a firm to foster strategic changes, both intra firm and inter firm (e.g. across supply chain), through building commitment to resource reconfiguration. It is critical that the culture of a firm fosters routines of sensing, seizing and reconfiguring.
2. Innovation in harnessing competence base: the skill set in a firm to deal with uncertain information (sensing) and develop business cases (seizing) incorporating substantial alterations to their business model as to deliver effective resource reconfiguration (reconfiguring).
3. Innovation in vision and strategy: ability of firms to rapidly develop and implement corporate strategies to enable them to engage with resource adoption and reconfiguration in a speedy manner. The strategic ability of a firm to rapidly seize and reconfigure a sensed opportunity is critical for competitive positioning.
4. Innovation in organisational intelligence: ability of a firm to blend 'planned' and 'experiential' approaches for iterative development of customer value propositions enabling firms to reconfigure resources to match market requirements. The approach to organisational intelligence in a firm is key in executing the routines of sensing, seizing and reconfiguring.
5. Innovation in idea management: ability of a firm to sense new ideas, seize them and reconfigure the resources to deliver the new idea, is key for success.
6. Integration of information systems: the ability of a firm to sense, seize and reconfigure and integrate new and existing ICT systems across the firm and its supply chain, is critical for business success.
7. Integration of strategy: ability of a firm to diligently couple e-business directions with corporate strategy directions to integrate resources. This is an extension of innovation in vision and strategy (item 3), but specifically reconfiguring e-technologies in the context of corporate strategy.
8. Integration of supply chains: ability of a firm to align new and existing channels to offer multi-channel operations for integrated distribution channels. Specifically, this refers to reconfiguring supply chains with electronic technologies. One could argue supply chain integration and ICT integration (Item 6) are closely aligned (see Donk, 2008; Johnson and Whang, 2002)

Therefore 'dynamic capabilities should be laid at the core of strategic management processes' (Shera and Lee, 2004: 935), wherein dynamic capabilities are tangible and intangible capabilities using resources effectively to deliver products and services. In essence, contextualising Teece's (2007) dynamic capability routines of sensing, seizing and reconfiguring through Daniel and Wilson's (2003) eight innovation and integrative capabilities, informs the potential of dynamic capabilities in the adoption of e-technologies.

Innovation and Integration in Supply Chains contextualised through Dynamic Capabilities

London (2008) identified that managing supply chains is about making improvements, particularly in: customer value, relationship management of trading partners, information management, flow of products and funds, competitiveness, innovation and reduction of costs. The supply chain context provides a lens through which to explore how trading partners are interconnected, particularly in terms of their goals, technologies, processes and relationships.

Siau and Tian (2004: 67) indicate that "the goal of supply chain integration is to link up the market place, the distribution network, the manufacturing process, and procurement activity in such a way that customers are better serviced at a lower total cost". Tan (2001: 44) suggests the goal of the integrated supply chain strategy is to create "manufacturing process and logistic functions seamlessly across the supply chain as an effective competitive weapon that cannot be easily duplicated by others", tantamount to dynamic capability. Therefore, sensing (innovation or integration) opportunities offered by e-commerce tools in the supply chain context especially global supply chain and seizing and reconfiguring resources enables firms to be agile and competitive in international markets.

Arcs of Integration by Frohlich and Westbrook (2001)	Description of Arcs of Integration by Frohlich and Westbrook (2001)	Types of Integrations by Fawcett and Magnan (2002)
(1) inward facing	Lower quartile for suppliers and lower quartile for customer	internal, cross-functional process integration
(2) periphery facing	Above lower quartile for suppliers or customers, but below upper quartile for suppliers and customers	
(3) supplier facing	In upper quartile for suppliers and below upper quartile for customers	backward integration with valued first-tier suppliers
(4) customer facing	In upper quartile for customers and below upper quartile for suppliers	forward integration with valued first-tier customers
(5) outward facing	In upper quartile for suppliers and in upper quartile for customers	complete forward and backward integration

Table 1 Integration by Frohlich and Westbrook (2001) and Fawcett and Magnan (2002)

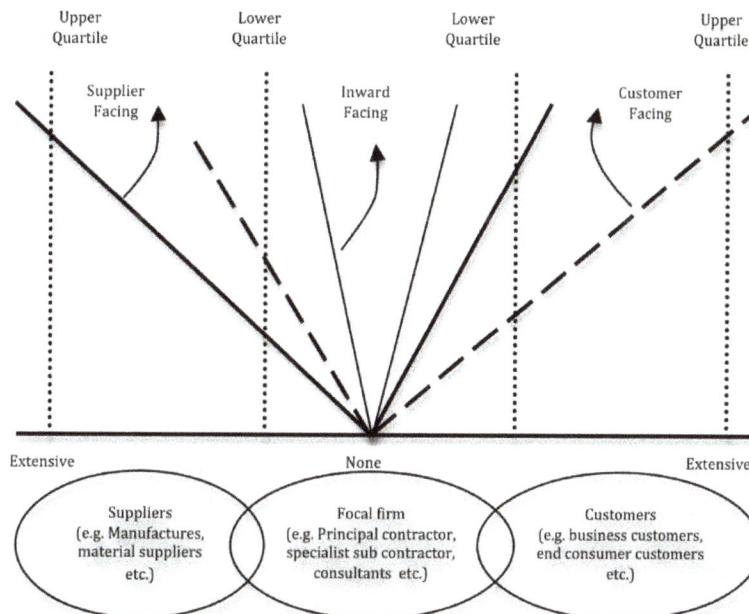

Figure 1 Arcs of integration modified from Frohlich and Westbrook (2001)

A firm's international business success partly depends on the extent innovation and integration is perpetuated by the members (suppliers and customers) and the functions (e.g. procurement, logistics, strategic planning etc.) of the supply chain. As indicated by Daniel and Wilson (2003), opportunities can arise from innovation in the culture, strategy, and management of ideas or organisation intelligence and/or from integration of information technology, strategy and supply chains. The extent of integration will depend on the contextual environment, including cultural, legislative, technical environments, within which the firms in the supply chain operate (Brisco and Dainty, 2005). Supply chain integration from a firm's perspective therefore, requires two modes of alignment: namely 'Information Integration' and 'Organisational Integration' (Bagchi and Skjoett-Larsen, 2003). Fawcett and Magnan (2002: 344) propose four types of integration, namely '(i) internal, cross-functional process integration; (ii) backward integration with valued first-tier suppliers leading to integration with second-tier; (iii) forward integration with valued first-tier customers; and (iv) complete forward and backward integration.' Extending the description of supply chain integration Frohlich and Westbrook (2001) identified five "arcs of integration" that describe the extent of integration across a supply chain using quartiles to position firms into one of the five categories (illustrated in Table 1 and Figure 1).

It is critical to note that the unique nature of the construction sector imposes additional layers of complexities, contributing to structural fragmentation, making supply chain integration more difficult in construction. The loosely coupled and transient nature of construction projects (Dubois and Gadde, 2002) manifests the construction project supply chains to be somewhat different to manufacturing supply chains. Each member of a project supply chain in construction e.g. consultant, project manager, contractor, sub contractors, suppliers and manufacturers, will be part of multiple transient supply chains concurrently. However, when firms develop long-term relationships leading to informal partnerships or consortia to deliver projects in a particular country and/or sector (e.g. project home, specialist healthcare, mining infrastructure etc.), such consortium's supply chains become relatively stable and less transient. This suggests that integration of a construction firm's project supply chains is complicated and firms in a project may not have total control of the entire project supply chain.

Consequently, individual firms need to develop dynamic capabilities to skilfully draw meaningful boundaries in their supply chains and create spheres within which they can foster and manage innovative initiatives, building up to sizeable competitive advantage. That is, each firm needs to skilfully strategise their supply chain integration at firm level and also on a project-by-project basis. As an example, a construction firm may have an overall strategy for a periphery facing supply chain, but in some projects they may choose backward integration while in other projects they may opt for forwarded integration as illustrated in Table 1 and Figure 1. Supply chain strategies can also be based on other perspectives, such as supply chain paradigms as outlined by Ayers (2002). He identified a number of paradigms including 'procurement, 'logistics and transportation', 'information' and 'strategic' as conceptualising supply chain operations. The paradigms are discussed below as contextualised in Daniel and Wilson's (2003) capability perspectives:

- The 'procurement paradigm' (PP) focuses on the procurement process across the supply chains and associate suppliers. The primary focus is improving the cost effectiveness of procurement process generally associated to the upstream supplier base. Sensing and seizing innovation and integration, the procurement approaches, specifically in the upstream spear (backward integration), and reconfiguring resources can iron out ineffectiveness in product or service delivery.

- The *'logistics and transportation paradigm'* (LP) is focused on physical movement of the products and services. This is a key paradigm for firms that manufacture and distribute products: that is, planning, implementing, controlling the efficient flow and storage of goods and services and related information from the point of origin to the point of consumption. Sensing and seizing innovation and integration in the transportation solutions is critical for effectiveness in product or service delivery. Both forward and backward integration in terms of logistics becomes appropriate, based on the nature of the business and the position of the firm in the construction project supply chain.

- The *'information paradigm'* (IP) focuses on improving information flows within the company and across the supply chain. This is assisted by integrated information systems, enabling effective flow of information to help improve coordination and cost of mistakes occurring due to lack of timely or accurate information. From the construction industry perspective, sensing and seizing innovation and integration of information sharing, has significant impact on effectiveness of business operations. Therefore, outward facing (backward and forward) integration can be argued to be appropriate.

- The *'strategic paradigm'* (SP) focuses on aligning the strategic goals to supply chain design and execution and has a long-term orientation. This has the view that innovation in vision and strategy is critical to improve the market share and profit while cost is secondary (Ayers 2002). Construction firms may focus on strategy integration among the firms within predetermined boundaries of a supply chain.

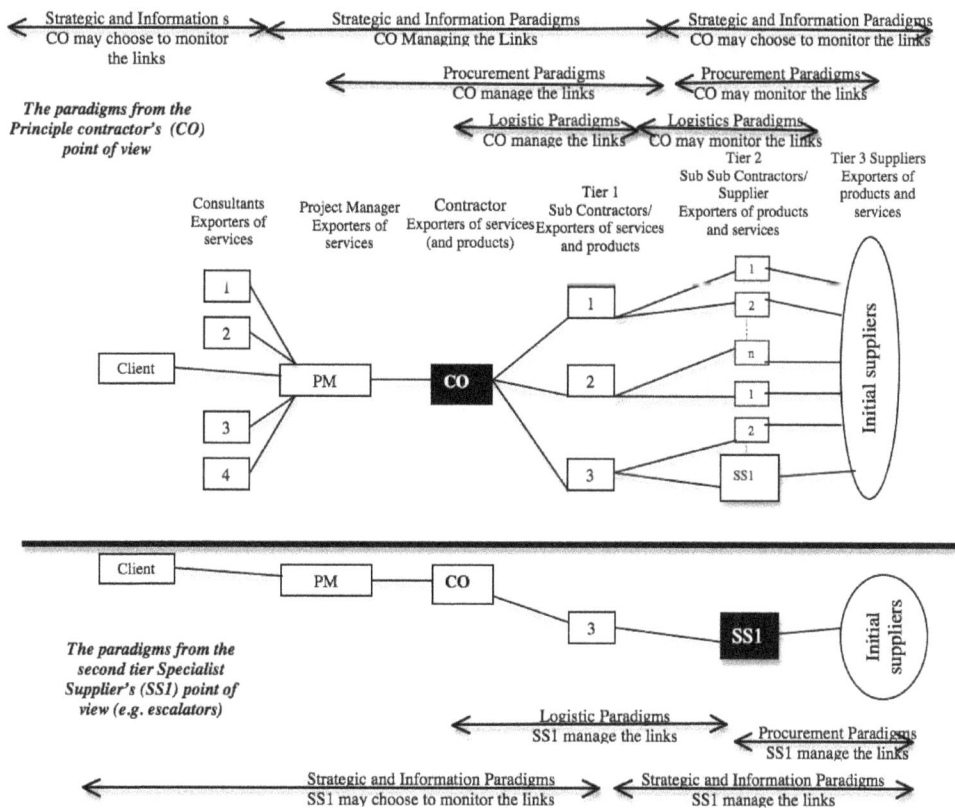

Figure 2 Supply paradigms from a contractor (top) and specialist supplier (bottom) perspectives

The above four paradigms are interconnected. Any successful supply chain design will encompass an appropriate mix of all the paradigms suitable in a selected business environment. Two examples on how to blend the different supply chain paradigms in a construction project context, focusing on a principal construction firm and a specialist supplier firm, are discussed below. Figure 2 presents a graphical representation of a construction project supply chain from the perspective 'principle contractor' and a 'specialist supplier' perspective (see Ash and Burn 2003 for a similar approach). Principal contractors roles in a project could differ based on the procurement method and contractual arrangement. The procurement or contractual differences along with other contextual factors, such as structure of the industry, transport and information infrastructure, in home and host countries, can influence the way each supply chain paradigm is exploited by a firm.

The supply chain in the top of Figure 2 identifies the dominant paradigms from a principal contractor's (CO) point of view hypothesised in a traditional procurement method, while the supply chain in the bottom focuses on a specialist supplier's (SS1) (e.g. escalator or lift firm) point of view. CO and SS1 can operate their upstream or suppliers' and downstream or customers' business activities in international markets. Hypothetically, an Australian contractor internationalising their construction business in Indonesia (either through export or FDI mode) essentially will need to appoint and manage a number of supply chain members including designers, sub contractors, suppliers etc. who have the possibility of operating from different geographical regions. For example, Leighton Asia, a subsidiary of Leighton Holdings (home country Australia), operating in Hong Kong, Macau, China, Mongolia, Taiwan, the Philippines, Thailand, Vietnam, Laos, Cambodia, Indonesia, Malaysia, Singapore and Brunei (Leighton, 2011) needs to manage their internationalised supply chains. As pointed out earlier, no firm including the CO, has control of their entire supply chain. Therefore drawing boundaries within the supply chains to identify the spears of activities links CO's need to 'manage' and 'monitor' (refer to Figure 2) and is crucial for effectiveness (Lambert and Cooper, 2000). The CO can use e-commerce tools to manage and monitor their supply chain activities in the international markets.

The supply chain representation of the Specialist Supplier (SS1) shares mostly the characteristics of manufacturing supply chains (Ayers, 2002). For example, OTIS (elevator business that will fit the description of SS1 in Figure 2) is a global company with local roots in more than 200 countries and territories. It has revenue of US \$11.7 billion (in 2009), of which 80 percent was generated outside its home country (the United States). Their major manufacturing facilities are in the Americas, Europe and Asia and engineering facilities are in the United States, Austria, Brazil, China, Czech Republic, France, Germany, India, Italy, Japan, Korea and Spain (OTIS, 2010). OTIS appears to have a globalised and regionalised international business and provides e-services to its customers. SS1 supply chain can be more integrated and more permanent than construction project supply chains, which can be fragmented and temporary. Most of the specialist suppliers in the construction industry could source their suppliers from different countries and manufacture in cost effective geographical locations, marketing the products internationally. Based on the nature of the business and supply chain design, a variety of e-commerce tools can be adopted by businesses to be competitive.

In summary, the framework along with the arcs of integration, can provide the basis to analyse the dynamic capabilities that can emerge from blending the four supply chain paradigms proposed by Ayers (2002).

Dynamic Capabilities and E-commerce

E-commerce exploits the digital networks to conduct business transactions by way of sharing business information and maintaining business relationships (Zwass, 2003). Moreover, e-commerce digitally enables commercial transactions between and among trading parties which involves exchange of value (e.g. money) across organisational or individual boundaries in return for products or services (Laudon and Traver, 2009). E-commerce is "the delivery of information, products or services, or payments via telephone lines, computer networks or any other means" (Kalakota and Whinston, 1996, p. 3) and serves "as a medium for enabling end-to-end business transactions" (Kauffman and Walden, 2001, p. 3). In essence, e-commerce uses intent and computer networking capabilities to perform business activities i.e. buy, sell or exchange products, services, and information (Turban et al., 2010). The principle objectives of e-commerce applications are to improve the efficiency of current practices and/or support the development of new practices (Johnson et al., 2002). E-commerce models help conduct traditional commerce through new ways of transferring and processing information, therefore providing the base for firms to develop dynamic capabilities.

The technological innovations arising from the combination of telecommunication and organisational computing shifted the directions of e-commerce from I-commerce (Internet Commerce) to M-commerce (Mobile commerce) (Wu and Hisa, 2008; Swilley, Hofacker and Lamont, 2012). Swilley, Hofacker and Lamont (2012) found in their study that, due to the growing necessity to gain competitive advantage, firms are ready to leverage capabilities gained from e-commerce into m-commerce. Embracing the opportunities arising from the shifts in e-commerce environments requires managers to constantly reconfigure their business resources-capabilities and meet emerging capability gaps in a timely manner (Zwass, 2003). Wu and Hisa (2008: 98) argue that e-commerce innovation can be attributed to a clever blend of technology with alternative business processes creating new forms of business models. A business model is a facilitating construct that blends the technologies and business values (Chesbrough and Rosenbloom, 2002) to provide structure to a business.

The application of e-commerce is not only about replacing paper trails or manual practices with digital alternatives but it implies more than that. E-commerce uses delicately interconnected electronic tools (e.g. computer networks, telephone, e-mail, electronic data interchange, internet, online collaborating tools and electronic funds transfer) to create a virtual network and virtual social space for trading partners or customers to communicate without requiring physical contact (Froehlich et al., 1999) making it attractive for international business. Corresponding to this, studies found that e-commerce can assist success of business internationalisation through relatively low cost business operations (Chai and Pavlou, 2004), enhancing the pace of business operations (Luo, Zhao and Du, 2005) and improving information and communication flow among all participants (Wang, Yang and Shen, 2007). E-commerce can reduce the cost of transactions for most businesses because it is easier to give the right offer to the right person at the right time, which in general, contributes to the efficiency of the business.

E-commerce can be classified into a number of modes including Business-to-Consumer (B2C), Business-to-Business (B2B) and Government to Business (G2B). Firms can engage with multiple e-commerce modes to form their international business. Major types of e-commerce are discussed below (Ash and Burn, 2003; Laudon and Traver, 2009; Turban et al., 2010; Dikbas and Scherer, 2004):

- Business-to-Consumer (B2C) e-commerce → online business selling to individual consumers. Most building supply firms directly reach the consumer. Moreover, the large residential builders (e.g. project home firms) also can directly deal with their customers with B2C tools.

- Business-to-Business (B2B) e-commerce → online business selling to other business. Most firms in a construction supply chain have the potential to engage with B2B systems. In general all suppliers and sub contractors can engage with the main contractor, project managers and beyond, making B2B a significant component of construction project business. Ash and Burn (2003) classifies B2B into a further two subsets B2B^s and B2B^c, where the latter is about a business dealing with another business which is a corporate customer (e.g. a project home builder) who then passes the products or services to end-customer (e.g. a house buyer).
- E-Government e-commerce → is when a government entity buys or sells goods, services or information from or to, a business (G2B). G2B in the context of the construction sector involves use of online e-government platforms to engage with construction approvals, payment etc. This reduces cost of transactions for firms, particularly the ones that are geographically distanced from government offices.
- Business-to-Employees (B2E) → subset of intra-business category in which the organisation delivers services, information, or products to individual employees. Large construction companies offer products and services to their employees via online platforms, e.g. gym memberships, training programs and payroll management etc.

In summary, it is proposed that sensing and seizing alternative e-commerce modes and reconfiguring organisational resources (i.e. dynamic capability framework) offers the opportunity in the design of the project supply chain enabling construction firms to develop competitive advantages (Shera and Lee, 2004). Fusing the four supply chain paradigms proposed by Ayers (2002) and innovative and integrative capabilities proposed by Daniel and Wilson (2003) within the dynamic capability framework enables a further level of meaningful organisational analysis.

Dynamic Capabilities embedded through the Coupling of Supply Chain and E-commerce Models

As indicated previously, firms that produce and sell their products and services with e-enabled supply chains via effective flow of material, plant, people, finances and information will position the firm with a significant competitive position (Lambert and Cooper, 2000). Specifically, from a construction industry point of view, the effective management of information is a key factor in improving quality, cost efficiency and shortened project delivery times. In information intense environments the implementation of a coherent supply strategy embedded in e-commerce is vital for competitive advantage (Dikbas and Scherer, 2004). Eight unique features of e-commerce technology identified by Laudon and Traver (2009) (see Table 2) reinforce the potential of the e-commerce technology to develop dynamic capabilities to assist construction firms to create competitive advantage in their international business. It is evident that the features of e-commerce, such as ubiquity, global reach and interactivity, assists in improving the supply chain operations of any business (Zhu and Kraemer, 2002; Lee, 2001; Swilley, Hofacker and Lamont, 2012), including international business. E-commerce can be embedded into all four-supply chain paradigms.

Figure 3 provides a graphical representation of e-commerce concepts applied to construction supply chains. Although Figure 3 is focused on depicting a contractor's supply chain, the e-commerce embedded supply chain framework can be used to analyse other firms. Firms in the supply chain can be from various countries operating in export mode or FDI mode. Each firm in a project based on a firm's core business and nature of inbound or outbound operations, will have a distinct supply network. This will impact on the dominant supply chain paradigm(s) and e-commerce tools aiding such paradigms. Supply chain connections in construction projects are underpinned by complex relationships that vary based on the procurement method, making it difficult to generalise atypical e-commerce approaches to internationalising the construction businesses. However, firms in the construction sector tend to be agile enough to cope with varying project supply chain needs. Table 2 identifies the key concepts that assist to align supply chain design to e-commerce tools.

E-commerce features (Laudon and Traver, 2009)		Construction internationalisation focus (Ayers, 2002)
Ubiquity – e-commerce technology is available everywhere (at home, at work, via mobile) at anytime (servicing 24 h a day, 7 days a week).	→	Ubiquity is critical for 'information paradigm'. This can allow firms located in home and host countries to work in different international time zones.
Global reach – the technology reaches across national boundaries around the world.		The global reach and universal standards of e-commerce can contribute to 'information and strategic' paradigms by enabling easy integration of information technology for effective information flow at strategic and operational levels.
Universal standards – create one set of technology standards (internet based) that is common, inexpensive, global technology foundation for business use.		
Interactivity – the technology works through interaction with the user. Consumers/suppliers are engaged in dialogues that dynamically adjust the experience to the specific requirements.	→	Interactivity can assist with 'procurement' paradigm by developing relationships with suppliers and reducing supply costs by sharing accurate information.
Richness – video, audio, text message are integrated into single message.		The rich, personalised and social nature of web based e-commerce applications can contribute to the 'strategic' paradigm, by way of enabling firms to develop relationships between suppliers and consumers, in home and host countries, to share complex business ideas.
Personalisation/Customisation – the technology allows personalised messages to be delivered based on individual or group characteristics.		
Social technology – the technology enables user content creation and distribution and supports social networks.		
Information density – the technology reduces information costs and raises quality. Information becomes plentiful, cheap and accurate.	→	Ability to deal with high information density can assist with 'logistic' and 'information' paradigms by providing accurate information during design and construction stages.

Table 2 E-commerce features and focus for internationalisation of construction business

Figure 3 (drawn using concepts borrowed from Lambert and Cooper 2000, Ayers 2002 and) proposes potential opportunities that could be sensed and seized that from fusing the supply chain concepts and electronic commerce concepts outlined by Lambert and Cooper (2000), Ayers (2002) and Ash and Burn (2003). This figure illustrates how principal contractors can use B2B platforms to manage the operations with tier 1 sub contractors and project managers. Firms can also use B2B platforms to assist in monitoring the operations of tier 2 firms and upstream members of the supply chains (e.g. consultants) with appropriate security permissions). Moreover, the contractors can use B2C platform to improve customer relationship management. Some of these operations can also be classified under B2B[c]. Although the engagement with G2B is driven by the government initiatives, firms exploiting the G2B opportunities can assist them with improving efficiencies.

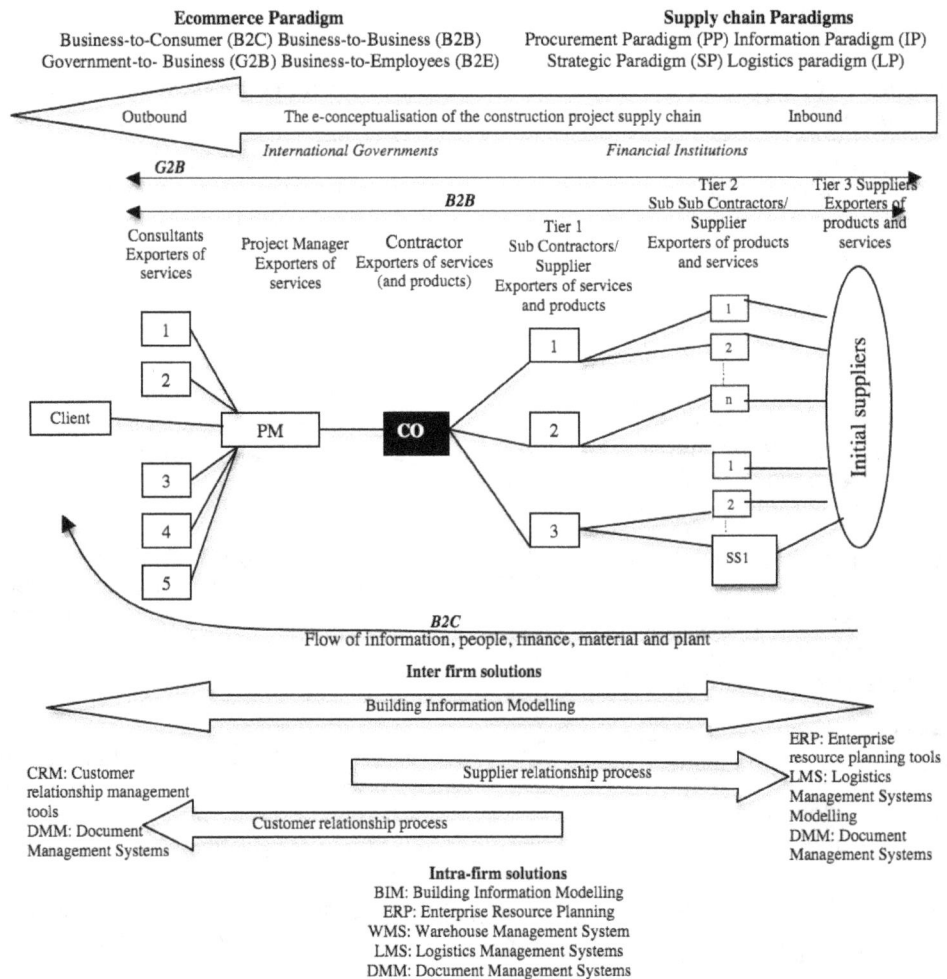

Ecommerce Paradigm
Business-to-Consumer (B2C) Business-to-Business (B2B)
Government-to- Business (G2B) Business-to-Employees (B2E)

Supply chain Paradigms
Procurement Paradigm (PP) Information Paradigm (IP)
Strategic Paradigm (SP) Logistics paradigm (LP)

Outbound   The e-conceptualisation of the construction project supply chain   Inbound

International Governments   *Financial Institutions*

G2B

B2B

Tier 2
Sub Sub Contractors/
Tier 3 Suppliers
Exporters of products and services

Consultants
Exporters of services

Project Manager
Exporters of services

Contractor
Exporters of services (and products)

Tier 1
Sub Contractors/
Supplier
Exporters of services and products

Supplier
Exporters of products and services

Client   PM   CO   Initial suppliers

1
2
3
4
5

1
2
3

1
2
n

1
2

SS1

B2C
Flow of information, people, finance, material and plant

Inter firm solutions
Building Information Modelling

CRM: Customer relationship management tools
DMM: Document Management Systems

Supplier relationship process

Customer relationship process

ERP: Enterprise resource planning tools
LMS: Logistics Management Systems Modelling
DMM: Document Management Systems

Intra-firm solutions
BIM: Building Information Modelling
ERP: Enterprise Resource Planning
WMS: Warehouse Management System
LMS: Logistics Management Systems
DMM: Document Management Systems

Figure 3 E-commerce in construction supply chains
(conceptualised from Lambert and Cooper 2000, Ayers 2002 and Ash and Burn, 2003)

Table 3 (constructed from concepts borrowed from, Ayers 2002 and Ash and Burn, 2003) proposes various broad categories of information technology systems that can assist to integrate different e-commerce initiatives aligned to supply chain-based initiatives (Issa, Flood and Caglasin, 2003). From a construction firm's point of view, G2B platforms will include any e-government platform set to deal with approval-related issues of construction projects. G2B may assist with information and procurement paradigms. The Customer Relationship Management Systems will be central for B2C initiatives, assisting with information and strategic paradigms. The B2B platforms can be focused on specific activities relating to the business. They can include Online Document Management (Alshawi and Ingirige, 2003), Enterprise Resource Planning Systems (Akkermans et al., 2003; Su and Yang, 2010), and Warehouse or Logistics Management Systems (Voordijk, Leuven and Lann, 2003). These systems can assist with all four paradigms from inter and intra firm perspectives. The Building Information Modelling (BIM) technology has the potential to impact on the information, strategic and procurement paradigms (McGraw Hill Construction 2009, 2010; 2012).

Paradigms	Type of E-commerce	IT/E-commerce Tools (Inter and Intra firm platforms)
Information Paradigm	G2B	e-Government portals (Inter)
	B2C	Customer Relationship Management Systems (Inter)
		Building Information Modelling (Inter & Intra)
	B2B	Online Document Management (Inter & Intra)
Strategic Paradigm		Enterprise Resource Planning Systems (Inter & Intra)
Procurement Paradigm		Warehouse Management Systems (Inter & Intra)
Logistics Paradigm		Logistics Management Systems (Inter & Intra)

Table 3 E-commerce in construction internationalisation of construction business

Primarily, contracting firms can exploit international markets by sensing and seizing opportunities offered by the e-commerce platform and reconfigure the firms' resources to (a) outsource business functions or operations to an international destination (upstream activities) and/or (b) offer the firm's products or services in the international market (downstream activities). The successful usage of e-commerce to support internationalisation is not without challenges. The asymmetry in e-commerce distribution seems to be caused not only by various levels of economic and socio-technical infrastructure, political and legal factors but also by cultural aspects in adopting e-commerce across nations. These have been recognised as major issues in the internationalisation of e-commerce (Kshetri, 2001). E-commerce can only be utilised competently at an optimum level if the employees and all clients have good ability to make use of the technology (Johnson and Whang, 2002). Thus, e-commerce platform for international business activities can be explored through dynamic capability routines exploiting innovation/integration capabilities.

Concluding Remarks

The construction firms can internationalise their business through import or export or FDI mode using an agile supply chain embedded with e-commerce capabilities. Exploiting international markets can occur in at least two forms: (a) outsource their selected business functions or operations to an international operator (supplier focus) and/or (b) offer the firm's products or services in the international market (customer focus). E-commerce allows firms, regardless of their size, type of business and geographical location to internationalise business in both FDI and import or export mode focusing on suppliers or customers.

In this paper it is proposed that fusing dynamic capability routines, namely sensing, seizing and re configuring proposed by Theece (2007) and 'Innovation' and 'Integration capabilities' proposed by Daniel and Wilson (2003) can provide an interesting analytical lens to explore adoption of emerging alterative ecommerce platforms in the context of a firms supply chain to gain competitive advantage. Dynamic capability framework enable on going sensing of (innovation and integration) opportunities, seizing the appropriate ones and re configuring resources to develop new business processes, products or models providing competitive advantage.

The paper specifically explored dynamic capability framework to analyse opportunities arising from innovation (in the culture, strategy, and management of ideas or organisation intelligence) and/or from integration (of information technology, strategy and supply chain) perpetuated by the members (suppliers and customers) in the functions (e.g. procurement, logistics, strategic planning etc.) of the supply chain. The discourse indicates that innovation in all forms assists supply chain integration namely; (i) internal, cross-functional process integration; (ii) backward integration with valued first-tier suppliers, leading to integration with second-tier; (iii) forward integration with valued first-tier customers; and (iv) complete forward and backward integration, is strongly coupled with ICT (electronic) platforms and tools.

In essence this paper proposes a framework to conceptualise internationalisation of construction business through conscious sensing and seizing e-commerce opportunities in the context of their supply chain and reconfiguring the firms' resources. Based on the dominant supply chain paradigm (strategic, information, procurement, logistics) underlying the firm's business model, firms can choose their e-commerce approach (B2B, B2C etc) and tools (CRM, ERP BIM etc). Firms also need to be conscious of the need for both effective inter-company wide systems and intra-company or supply chain wide systems. It proposes that dynamic capabilities enable firms to exploit e-commerce platforms to channel upstream activities to an international destination and explore the opportunities to offer products and services to the international market.

References

Akkermans, H. A. Bogerd, P. Yucesan, E. and van Wassenhove, L. N. (2003) The impact of ERP on supply chain management: Exploratory findings from a European Delphi study, *European Journal of Operational Research*, **146**, 284-301

Alshawi, M. and Ingirige, B. (2003) Web-enabled project management: an emerging paradigm in construction, *Automation in Construction*, **12**, 349-364

Ambrosini, V and Bowman, C (2009) What are dynamic capabilities and are they a useful construct in strategic management? *International Journal of Management Reviews*, **11** (1), 29-49

Ash, C. G. and Burn, J. M. (2003) Assessing the benefits from e-business transformation through effective enterprise management. *European Journal of Information Systems*, 12, pp. 297-308.

Ayers, J. B., Ed. (2002) *Making Supply Chain Management Work*, Auerbach Florida

Bagchi, P. K. and Skjoett-Larsen, T. (2003) Integration of Information technology and organisations in a supply chain, *The International Journal of Logistics Management*, **14** (1), 89-108

Barreto, I, (2010) Dynamic Capabilities: A Review of Past Research and an Agenda for the Future, *Journal of Management,* **36** (1), 256-280

Briscoe, G. and Dainty, A. (2005) Construction supply chain integration: an elusive goal? *Supply Chain Management*, **10** (4), 319-325

Chai, L. and Pavlou, P. (2004) From 'ancient' to 'modern': a cross-cultural investigation of in electronic commerce adoption in Greece, *Journal of Enterprise Information Management*, **17** (6), 416-23

Chesbrough, H. and R. S. Rosenbloom (2002) The role of the business model in capturing value from innovation: Evidence from Xerox Corporation's technology spin-off companies, *Industrial and Corporate Change*, **11** (3), 529-555

Cutting-Decelle, A-F., Young, B. I., Das, B. P., Case, K., Rahimifard, S., Anumba, C. J. and Bouchlaghem, D. M. (2007) A review of approaches to supply chain communications: from manufacturing to construction, *ITcon*, **12** (2007), 73-102

Daniel, E. M. and Wilson, H. N. (2003) The role of dynamic capabilities in e-business transformation, *European Journal of Information Systems*, **12**, 282-296

Dikbas, A. and Scherer, R., Eds. (2004) *eWork and eBusiness in Architecture Engineering and Construction* ECPPM, Taylor & Francis, London.

Donk, D. P. v. (2008) Challenges in relating supply chain management and information and communication technology: An introduction, *International Journal of operations & production management*, **28** (4), 308-312

Dubois, A. and Gadde, L.-E. (2002) The construction industry as a loosely coupled system: implications for productivity and innovation, *Construction Management and Economics*, **20** (7), 621-631

Eisenhardt, K. M. and Martin, J. A. (2000) Dynamic capabilities: What are they? *Strategic Management Journal*, **21** (10/11), 1105-1121

Eyob, E. and Tetteh, G. (2012) *Customer-Oriented Global Supply Chains: Concepts for Effective Management*, (pp. 0-335), IGI Global, Web. 22 Mar. 2012. doi:10.4018/978-1-4666-0246-5

Fawcett, S. E. and Magnan, G. M. (2002) The rhetoric and reality of supply chain integration, *International Journal of Physical Distribution and Logistics Management*, **35** (5), pp. 339-361

Fawcett, S. E. Osterhaus, P. Magnan, G. M. Brau, J. C. and McCarter, M. W. (2007) Information sharing and supply chain performance: the role of connectivity and willingness, *Supply Chain Management: An international Journal*, **12** (5), 358-368

Froehlich, G., Hoover, H.J., Liew, W. and Sorenson, P.G. (1999) Application framework issues when evolving business applications for electronic commerce, *Information Systems*, **24** (6), 457-473

Frohlich, M. T. and Westbrook, R. (2001) Arcs of integration: an internal study of supply chain strategies, *Journal of Operations Management*, **19**, 185-200

Howes, R. and Tah, J. H. M. (2003) *Strategic Management Applied to International Construction*, Thomas Telford, Victoria

Issa, R. R. Flood, I. and Caglasin, G. (2003) A survey of e-business implementation in the US construction industry, *ITcon*, **8**, 15-28

Johnson, E and Whang, S (2002) E-business and supply chain management: An overview and framework, *Production and Operations Management*, **11** (4), 413-23

Johnson, R., Clayton, M., Xia, G., Woo, J.H. and Song, Y. (2002) The strategic implications of e-commerce for the design and construction industry, *Engineering Construction and Architectural Management*, **9** (3), 241-248

Kalakota, R. and Whinston, A.B (1996) *Frontiers of electronic commerce*, Addison Wesley Longman Publishing Co., Inc. Redwood City, CA, USA

Kauffman, R.J. and Walden, E.A. (2001) Economics and electronic commerce: Survey and directions for research, *International Journal of Electronic Commerce*, **5**, 5-116

Kshetri, N.B. (2001) Determinants of the locus of global e-commerce, *Electronic Markets*, **11** (4), 250-257

Lambert, D. M. and Cooper, M. C. (2000) Issues in Supply Chain Management, *Industrial Marketing Management*, **69** (1), 65-83

Laudon, K C. and Traver, C. G. (2009) *E-Commerce: Business, Technology, and Society*, 2nd. Ed., Addison Wesley

Lee, C-S (2001) An analytical framework for looking at e-commerce business models and strategies. *Internet Research: Electronic Networks, Applications and Policy*, **11** (4), 349-59

Leighton Group. (2011). "About us", at http://www.leightonasia.com/v4/default.asp?lid=1&sec=About+Us (accessed May, 2011)

London, K. (2008), *Construction supply chain economics*, Taylor & Francis, London

London, K. (2010). Multi-market industrial organisational economic models for the internationalisation process by small and medium enterprise construction design service firms, *Architectural Engineering and Design Management,* **6** (2), 132-152

Luo, Y., Zhao, J.H. and Du, J. (2005) The internationalisation speed of e-commerce companies: an empirical analysis, *International Marketing Review*, **22** (6), 693-709

McGraw Hill Construction (2009) The business value of BIM: Getting building information modelling to the bottom line, In: *Smart Market Report*

McGraw Hill Construction (2010) The business value of BIM in Europe: Getting building information modelling to the bottom line the united kingdom, France and Germany. In: *Smart Market Report*

McGraw Hill Construction (2012) *The business value of BIM for infrastructure: Addressing America's infrastructure challenges with collaboration and technology*

Makadok, R. (2001) Toward a synthesis of the resource-based and dynamic-capability views of rent creation, *Strategic Management Journal*, **22** (5), 387- 401

Menipaz, E. and Menipaz, A. (2011) *International Business*, SAGE, London

Min, H. and Zhou, G. (2002) Supply chain modelling: past, present and future. *Computers & Industrial Engineering*, 43, 231-249

Nelson, R R and Winter, S G (1982) *An evolutionary theory of economics*, Cambridge: Harvard University Press.

O'Reilly III, C A and Tushman, M L (2008) Ambidexterity as a dynamic capability: Resolving the innovator's dilemma, *Research in Organizational Behavior*, **28** (July), 185-206

OTIS. (2010). *OTIS Worldwide*, at http://www.otisworldwide.com/ (accessed May 2011, 2011)

Pant, S. Sethia, R. and Bhandarib, M. (2003) Making sense of the e-supply chain landscape: an implementation framework, *International Journal of Information Management*, **23** (3), 2001-221

Porter, M E (1980) *Competitive strategy: Techniques for analyzing industries and competitors*, New York: Free Press.

Porter, M.E. (1990) *The Competitive Advantage of Nations*, Free Press, New York

Reina, P. and Tulacz, G. J. (2010a) *The top 200 international design firms*, ENR, ENR, New York

Reina, P. and Tulacz, G. J. (2010b) *The top 225 international contractors*, ENR, ENR, New York

Rindova, V. P. and Kotha, S. (2001) "Morphing": Competing through Dynamic Capabilities, Form, and Function, *The Academy of Management Journal*, **44** (6), 1263-1280

Roy, S. Sivakumar, K. and Wilkinson, I. F. (2004) Innovation Generation in Supply Chain Relationships: A Conceptual Model and Research Propositions, *Journal of the Academy of Marketing Science*, **32** (1), 61-79

Rugman, A. M. and Collinson, S. (2009) *International Business*. Prentice Hall, Sydney

Shera, P. J. and Lee, V. C. (2004) Information technology as a facilitator for enhancing dynamic capabilities through knowledge management, *Information & Management*, **41**, 933-945

Siau, K. and Tian, Y. (2004) Supply chain integration: architecture and enabling technologies, *The Journal of Computer Information Systems*, **44** (3), 67-72

Su, Y. and Yang, C. (2010) A structural equation model for analyzing the impact of ERP on SCM. *Expert Systems with Applications*, **37**, 456-496

Smart, A. (2008) eBusiness and supply chain integration, *Journal of Enterprise Information Management*, **21** (3), 227-246

Swilley, E., Hofacker, C. F., and Lamont, B.T (2012) The Evolution from E-Commerce to M-Commerce: Pressures, Firm Capabilities and Competitive Advantage in Strategic Decision Making, *The international Journal of e-Business Research*, **8** (1), 1-16

Tan, K. C. (2001) A framework of supply chain management literature, *European Journal of Purchasing and Supply Management*, **7**, 39-48

Teece, D. and Pisano, G. (1994) *The Dynamic Capabilities of firms: An introduction*, The International Institute for Applied Systems Analysis, Austria

Teece, D. J. (2007) Explicating dynamic capabilities: The nature and microfoundations of (sustainable) enterprise performance, *Strategic Management Journal*, **28**, 1319-1350

Teece, D. J. Pisano, G. and Shuen, A. (1997) Dynamic capabilities and strategic management, *Strategic Management Journal*, **18** (7), 509-533

Turban, E, Lee, J K, King, D, Liang, T P and Turban, D (2010) *Electronic commerce*, 6th ed. New Jersy: Prentice Hall Press

Voordijk, H. Leuven, A. V. and Lann, A. (2003) Enterprise Resource Planning in a large construction firms: implementation analysis, *Construction Management and Economics*, **21**, 511-521

Vrijhoef, R. and Koskela, L. (2000) The four roles of supply chain management in construction, *European Journal of Purchasing & Supply Management*, **6** (2000), 169-178

Wang, Y., Yang, J. and Shen, Q. (2007) The application of electronic commerce and information integration in the construction, *International Journal of Project Management*, **25** (2), 158-163

Winter, S. G. (2003) Understanding dynamic capabilities. *Strategic Management Journal*, **24**, 991-995

Wu, J. H. and Hisa, T. L. (2008) Developing E-Business Dynamic Capabilities: An Analysis Of E-Commerce Innovation From I-, M-, To U-Commerce, *Journal of Organisational Computing and Electronic Commerce*, **18**, 95-111

Zhu, K and Kraemer, K L (2002) E-commerce metrics for net-enhanced organizations: Assessing the value of e-commerce to firm performance in the manufacturing sector, *Information Systems Research*, **13** (3), 275-95

Zwass, V. (1996) Electronic commerce: structures and issues, *International Journal of Electronic Commerce*, **1** (1), 3-23

Zwass, V. (2003) Electronic commerce and organisational innovation: aspects and opportunities, *International Journal of Electronic Commerce*, **7** (3), 7-37

---

[i] International design: building (US$8,504.6 Million [M]), industrial/petroleum (US$ 21,351.0M), manufacturing (US$ 404.1M), transportation (US$8,985.5M), power (US$4,469.6M), hazardous waste (US$ 2,090.4M), water (US$ 2,755.7M) and sewer/waste (US$ 1,762.8M) (values are for the year 2010-Source Reina and Tulacz, 2010a).

[ii] International contracting: building (US$ 86.0 Billion [B]), industrial /petroleum (US$ 112.0B), manufacturing (US$ $3.8B), transportation (US$ $112.3B), power (US$ 35.7B), hazardous waste (US$ 0.8B), water (US$ 11.2B), Telecommunication (US$ 2.7B) and sewer/waste (US$ 6.3B) (values are for the year 2010 - source, Reina and Tulacz, 2010b).

Framework for Structuring Procurement Contracts 5

Lena Borg & Hans Lind

Royal Institute of Technology, Sweden

Abstract

The aim of this paper is to propose a new framework for structuring contract types and payment methods. Concerning procurement contracts, the first important new feature of this framework is a stepwise structure with three main steps in the contract design: (1) what will be procured—should the contract only include construction, or should it include both construction and operation/maintenance (2) who will do the detailed design of the premise and (3) how many contractors will the client use? The second important new feature of this framework is that both step 2 and step 3 include a continuum of alternatives. Concerning payment methods, the new framework is primarily based on how the specific risks of the project are shared. These frameworks can be useful for policy formulation in that they can help to avoid some problematic ways of formulating policies.

Keywords: Procurement, Contract, Construction sector, Infrastructure projects

Introduction

Discussions about productivity problems and cost overruns are common in many countries. Changes in procurement have been proposed as ways to create incentives for innovation and for taking life-cycle cost into account (Mandell & Nilsson 2010). These changes involve moving from Design-Bid-Build (DBB) contracts to Design-Build (DB) contracts, and/or to contracts in which construction and maintenance are bundled, such as in Public Private Partnership (PPP).

A survey done by Eriksson and Laan (2007) shows that for the majority of projects procured as DBB contracts, the clients and their consultants make the detailed design together. In these cases, it is possible for the design to be handled in-house if clients use their own staff (SOU 2009:24). On the other hand, in a typical DB contract the client specifies the general characteristics of the end product. This can theoretically be done in a number of ways, such as: by referring to earlier products ('we want a standard type of this'); by specifying the general characteristics of the house ('we want a residential building in seven floors with x square metres and fulfilling basic legal quality demands'); or by specifying various functional characteristics of the object (e.g., Bejrum & Grennberg 2003; Mattsson & Lind 2009). The fundamental difference between DBB contracts and DB contracts is who has the responsibility for the detailed design; in the first, the responsibility lies with the client, while in the second, the responsibility lies with the contractor. However, in both cases, the client typically has the responsibility for the operation and maintenance phases. In the construction of some projects, such as PPP projects, operation and maintenance is bundled to one contract (Leiringer 2003; Lind & Borg 2010). It is argued that this kind of contract (both DB contracts and PPP) gives the contractor a higher degree of freedom and the ability to use new solutions to cut costs and resources (Ng & Wong 2007).

The starting point for this paper is the belief that logical and clear terminologies and clearly structured arguments are important in a number of contexts. Clients have to make decisions about how to procure projects and if the alternatives are described in a vague and unsystematic way, then there is a risk for incorrect decisions, and the optimal procurement contract is not chosen. The framework determines how we formulate the alternatives and how we think about an issue. Clarity and logic are also important from a scientific perspective. If you want to compare and evaluate procurement contracts and find out their advantages and disadvantages, it is important that the alternatives are described in a logical and clear terminology. Otherwise we will not know what has been compared and what characteristics of the contract are responsible for the observed consequences. Without well-structured and clear alternatives it will be difficult to draw policy implications from research.

Our aim is to present a new and simple framework for describing and analysing alternative procurement and payment systems. As shown below, the definitions in leading textbooks lack consistency. In this article, we focus on contracts for infrastructure projects, such as roads and railways, which typically have a public client. This is a conceptual paper based on a selective sample of literature. The books we discuss here were chosen because they are leading textbooks in construction management[1]. The structure of the paper is as follows. In the next section, we examine how contracts and payment forms are described and categorised in leading textbooks (Gould & Joyce 2011; Ritz & Levy 2013; Winch 2010). Then we present our proposed framework, followed by some general reflections about the choice of procurement contract and payment methods. In the subsequent section, we present reflections on the choice of contract type and payment mechanism. In the final section, we present our general conclusions and the advantages of the proposed framework will be clarified.

How Contracts are Structured in Selected Literature

It is common in the selected literature to start with a rather long list of procurement contract types without a clear system: DBB contracts, DB contracts, performance-based contracts, PPP contracts and more. Each contract is seen as a unique entity with specific characteristics. Types of contracts are often graded in terms of additional commitment for the contractor. Secondly, the selected literature contains no common terminology for the whole problem at hand. American literature primarily uses the term *Project Delivery Method* (see Gould & Joyce 2011; FHWA n.d.) while Winch (2010) uses the term *ways of procuring*. Ritz and Levy (2013) use the term *contract executing approach*.

Contract type refers to payment method in Gould and Joyce (2011); but payment method is called *Contract format* in Ritz and Levy (2013). In Federal Highway Administration (FHWA) publications, *Procurement Method* refers to the selection criteria used when choosing a contractor (FHWA n.d.), which are called *Ways of procuring* in Winch (2010).

We recommend using the basic terminology *Procurement contract type* and *Payment method*. The first term refers to how tasks are allocated between different actors, and the second term refers to how the contractor is paid.

Procurement Contract Type

The tables below, and the comments after the tables, summarise how procurement contract types are structured in different sources. We first look at Gould and Joyce (2011), as shown in Table 1. In the text, Gould and Joyce (2011) also discuss:

- Concession contracting (p. 34) including DBFO (Design-Build-Finance-Operate) and BOOT (Build-Own-Operate-Transfer); and

---

[1] Based on interviews with Swedish lecturers in the area.

- Innovation in project delivery (p. 91), where they mention PPP as a way to finance and give BOO (Build-Own-Operate), DBO (Design-Build-Operate), and DBF (Design-Build-Finance) as further examples.

Table 1: Contracts structure in Gould and Joyce (2011, Ch. 4)

Name of contract type	Description
Design-Bid-Build	The client hires a designer (architect), who prepares a design and completes contract documents. With correct documents, the client either conducts a bidding process or negotiates with a specific contractor. The contractor is then responsible for constructing and delivering a complete project. Both the architect and the contractor have the option of choosing subcontractors. The contractor is solely responsible for the execution of the work.
Design-Build	The client hires a firm, that is, a contractor that will perform both design and construction. The contractor has the option of hiring subcontractors and architects for the design.
Construction Management	The client hires both a construction management firm and a designer (architect) and has the sole responsibility of hiring individual construction contractors. The construction manager can vary in expertise and can be put in place at different stages. The architect is free to hire subcontractors.

Table 2 describes the structure presented in Winch (2010), who uses the term *formation of the project coalition*. Winch (2010) also discusses four basic types of privately financed procurements (p. 43): Concession, Private Finance Initiative, Public Private Partnerships, and Company Limited with Guarantee.

Table 2: Contract structure in Winch (2010, Ch. 5)

Name of contract type	Description
Separated	The client hires suppliers and (designer) architects and uses competitive tendering to obtain subcontractors. The architect is then responsible for selecting the trade contractors who will execute the site work. The architect is responsible for co-ordinating the contractors, but is not responsible for any failings on their part. One version involves the client hiring a general contractor that takes over the responsibility of the execution of the project on-site.
Integrated (Turnkey)	The client hires a single contractor for both the design and construction stages on a competitive tender basis.
Mediated (construction project manager)	The client hires architects, as well as a construction manager who is responsible for managing the trade contractors on site. The contractors are selected on the basis of a competitive tender organised by the construction manager. The arrangements and terminology vary considerably depending on the clients' or the construction managers' different responsibilities at various stages.
Unmediated	The client has a high level of in-house project management capability, and has the necessary knowledge for and option of hiring subcontractors.

Table 3 presents the contract structure in Ritz and Levy (2013). In the text, Ritz and Levy (2013, p. 51) also mention Build-Operate-Transfer (BOT) as another alternative. The FHWA (n.d.) presents the structure outlined in Table 4.

Table 3: Contract structure in Ritz and Levy (2013, Ch. 2)

Name of contract type	Description
Traditional	The client hires a separate designer and a single general contractor that both have the option of hiring subcontractors.
Turnkey	The client has two options. The first option is design-build, where a single engineering contractor has the responsibility for both the design and construction. The general contractors hired by the engineering contractor have the option of either hiring subcontractors or using their own workforces. In the second option, the client hires an engineering construction manager with the responsibility for the design and construction, who in turn hires a designer and a construction manager who have the responsibility for the construction and for possible subcontractors.
Owner Builder	The client is responsible for design and construction and has the option of either using in-house competence or hiring subcontractors.
Construction Management	The client has two options: first, hiring a separate designer and a general contractor that acts both as a construction manager and as the client's agent, with both designer and contractor having their own hired subcontractors; or second, hiring a designer, a construction manager that acts as the client's agent, and individual construction contractors.

Table 4: Contract structure in FHWA (n.d.)

Name of contract type	Description
Design-Bid-Build	The client hires separately for design and construction services, and keeps a high level of both control and risk. The contractor's involvement is restricted to the construction phase. The client completely defines the scope.
Design-Build	The client combines design and construction under a single contract. The contract can also cover design-build-maintain, design-build-warranty and design-build-operate. The owner has the option of defining a scope of work, but has opportunity for innovation. This type of contract is often used for projects that are complex in nature or have a high level of urgency. The contractor's involvement runs from just after the pre-design and ends at least after the warranty have expired.
Construction Management	The client hires a construction manager to act as a construction advisor during the pre-construction phase and as a general contractor during the construction. This contract transfers the cost and risk to the construction manager. The client has control over the scope and design during the process. The contractor's involvement runs from just after the pre-design and ends when the warranty has expired.
Public Private Partnership (PPP)	The client hires a developer who takes part in the financing of the project in return for the ability to collect toll revenues, or to pursue development rights. The developer is responsible for the integrated delivery of design, construction, and operation and maintenance for a time period specified in advance.
Alliance contracting/ Integrated project delivery (IPD)	The client and at least one service provider, such as constructors, consultants, and designers, collaborate on the delivery of a project. The client collaborates with the industry to allocate risk.

It can be seen that three of the procurement contract types are repeated in several of the classifications. *Design-Bid-Build* is explicitly present in two of the four classifications, and appears to be the same as what is called *Separated* or *Traditional* in the remaining two. *Design-Build* is also present in two of the tables and is called *Integrated* or *Turnkey* in the other two. *Construction Management* is present in three of the tables. What Winch (2010) calls *Unmediated* seems to be similar to what Ritz and Levy (2013) call *Owner Builder*. Several of the books include some form of PPP as a fourth alternative, although others see PPP as being outside their classification system, and merely comment on it in their text.

Payment Methods

Concerning payment methods, Tables 5, 6 and 7 summarise the main alternatives, as described by the authors. Again, there is a lack of common structure in these classifications, even though some forms are repeated a number of times. *Fixed price* is mentioned by all three classifications, and *Cost-plus* and *Unit price* contracts are mentioned by two. In addition, there appears to be no common structure in how alternatives are presented.

Table 5: Payment method structure in Gould & Joyce (2011, Ch. 4)

Name of payment method	Description
Single fixed price	Also called lump sum, this is a contract in which the contractor has agreed to deliver a specified amount of work for a specific sum of money. Once the contract is signed, both parties have to live with the terms.
Unit price contract	The client and contractors agree on the price that will be charged per unit for the major elements. The client often provides estimated quantities, and the contractor calculates the final price according to this information, with additions for the contractors' overhead, profit, and other project expenses. The final contract price is not known until the final work has been done.
Cost plus a fee	This is a contract in which the contractor is reimbursed by the client for all work costs, and also receives an additional agreed-upon fee, or a fee that is a percentage of the costs.

Table 6: Payment method structure in Winch (2010, Ch. 6)

Name of payment method	Description
Fee based	This cost-reimbursable contract also seems to cover a unit price contract.
Incentive contract	This can be both a fee based and a lump sum contract, and varies in outline. The consistent part of this type of contract is the attempt to have positive incentives within the contract, to motivate performance fulfilment by gainsharing between parties.
Fixed price	A contract in which the price is fixed for an agreed-on amount of work. It may be that the contractor's price is fixed, or it may be an after-measurement, if the quantity of work to be done is not known in advance.

Table 7: Payment method structure in Ritz and Levy (2013, Ch. 2)[2]

Name of payment method	Description
Cost-plus (a number of versions)[3]	The client agrees to pay the contractors for the cost of the work plus a fee, very often calculated as a percentage of the cost. This contract can be complemented with a guaranteed maximum, a guaranteed maximum and incentive, or a guaranteed maximum and provision for escalation.
Bonus (a number of versions)	The bonus in this type of contract may be related, for example, to time, completion, and/or performance.
Lump sum (a number of versions)	A contract in which contractors prepare their bids according to a completed set of plans and specifications. No more and no less than is stipulated in the documents should be included.
Unit price contracts	The client and contractors agree on the price that will be charged per unit for the major elements. The client often provides estimated quantities, and the contractor calculates the final price according to this information, with additions for the contractors' overhead, profit, and other project expenses. The final contract price is not known until the final work has been done.

The Proposed Basic Framework: Procurement Contract Type

We believe that similarities and differences between procurement contract types become clearer if a stepwise procedure is used; that is, a structure in which one dimension is introduced at a time. The following framework is based on three steps: determining what is to be procured; determining who will do the design; and determining how many contractors will be used.

Step 1: What Is to Be Procured; Construction Only, or Construction with Operating/Maintenance?

It is confusing that, especially in the FHWA (n.d.) framework, 'delivery methods' do not concern different ways of 'delivering' the same type of object. In addition, the composition of the object differs between methods. In one method, 'delivery methods' only concern a 'premise' - for example, a road or a tunnel - while in another method, they concern both building an object *and* operating/maintaining it for a considerable number of years.

In our proposed framework, the *first* step for the client is to decide whether a contract that delivers an object should be chosen, or whether a bundled contract that includes both construction and operation/maintenance should be chosen. The descriptions above show that PPP and BOT projects are not integrated in their basic framework. Instead, these types of contracts are mentioned in the text without a clear relation to their basic framework. Here, they are integrated in the same framework as traditional contracts that only concern a premise. Figure 1 illustrates this first step.

---

[2] They also discuss convertible contracts used in joint ventures, which is not relevant here.
[3] Here we also include what they call *Time* and *materials*.

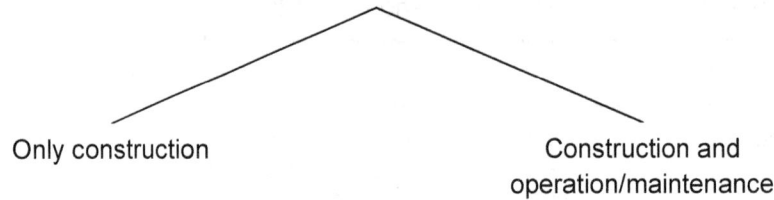

Figure 1: Initial decision when producing a contract in the infrastructure sector (Source: authors)

Step 2, Version 1: Who Will Do the Design?

The line drawn between DBB and DB contracts concerns who is responsible for the detailed designs of the facility. The same distinction is drawn between what Winch (2010) calls *Separated* versus *Integrated* contracts, and what Ritz and Levy (2013) call *Traditional* versus *Turnkey* contracts. In the first type of contract, the detailed design is the client's responsibility. In the second type, the detailed design is the contractor's task. This distinction is illustrated in Figure 2.

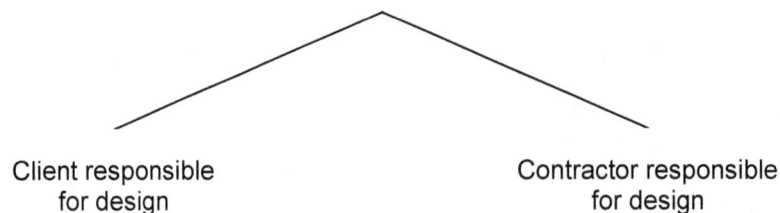

Figure 2: Step 2, Version 1: Who will do the design? (Source: authors)

In the literature, the distinction related to who makes the design is only used for pure construction contracts. However, the same distinction can be made for bundled contracts also. The client may have a clear view of exactly the kind of facility they want and how it should be managed, but may still write a bundled contract. The study presented by Borg (2011) indicates that in the (few) Swedish PPP projects that have been carried out, there was very little innovation. To a large extent, the contractor in these cases used techniques that the client had used earlier in DBB contracts. The choice of bundling construction and operation/maintenance can be motivated by arguments other than giving the contractor freedom concerning the design. For example, efficiency in the operation of the facility can motivate a bundled contract. Combining Figure 1 and Figure 2 therefore gives four basic options; however, as we argue in the next section, the real world options do not fit neatly into this framework.

Step 2, Version 2: Who Will Do the Design?

Nyström, Lind and Nilsson (2014) show that one cannot assume that a so-called 'DBB contract' has fewer degrees of freedom for the client than a so-called 'DB contract'. They also make it clear that most DB contracts include detailed technical specifications concerning a number of aspects of the premises. In order to simplify repairs or the handling of spare parts, the client might have very specific demands concerning some components, while leaving other things open. In practice, the responsibility for the detailed design is divided between the client and the contractor. It is therefore more correct to talk about a continuum of contract forms than to talk about just two alternatives (i.e., the client versus the contractor being responsible for the design). This continuum is illustrated in Figure 3, with an arc between the two extreme points. In one extreme, the client makes detailed design choices for all components; in the other extreme, the client only formulates rather general functional demands (e.g., the capacity of a road, maximum track depths). Each point on the arc represents a specific division of responsibilities for the detailed design.

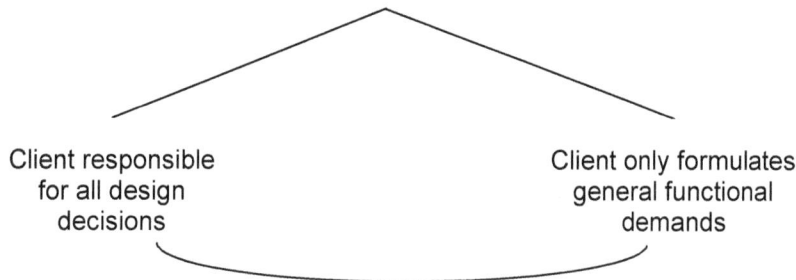

Client responsible   Client only formulates
for all design   general functional
decisions   demands

Figure 3: Step 2, Version 2: Who will do the design? (Source: authors)

If the client is responsible for design decisions, the framework contains a further subdivision concerning whether the design department is in-house, or whether independent consultancy firms are contracted for the design work. For example, the Swedish Transport Authority (STA) has gone from an in-house design department to almost complete outsourcing over the last fifteen years. This subdivision could be added to as 'step 2b' in the diagram. We do not include it here, in order to avoid unnecessary complexity in Figure 3.

External or In-house Project Manager: Construction Management

In the American literature in particular, Construction Management (CM) is described as one of the basic procurement strategies. In Sweden this is not seen as a specific 'delivery method' or contract form for infrastructure procurement (Eriksson & Hane 2014), but as a more pragmatic issue of whether to have an in-house project manager or whether to hire an external project manager. The STA, for example, sometimes uses a combination of CM and both external and internal project managers within the same project.

In the literature, CM is sometimes described as having a role in which the construction manager is almost the same as a contractor. The construction manager is described as being responsible to the client, and the construction manager hires subcontractors. Large contractors in Sweden, such as Skanska, PEAB, and NCC, currently describe themselves as CM companies, because they use subcontractors to a large extent. In this way, these contractors can reduce their fixed costs and risks. The comparative advantage for the company is being able to put together the right team of companies for a specific task.

In our proposed framework, CM is for these reasons not seen as a specific procurement contract type. How the contractor structures their work is up to them, and is not part of the procurement contract type.

Step 3: How Many Contractors Will Be Used?

In the models discussed above, it has been assumed that there is only one 'general' contractor, but this is of course not necessary. In the literature, there were models like the one Winch (2010) calls *Unmediated* or what Ritz and Levy (2013) call *Owner Builder*, in which the client hires several contractors to carry out specific tasks. Our framework therefore includes a continuous scale concerning the number of subcontractors: from one general contractor to a large number of separate subcontractors. This scale is shown in Figure 4.

It is important to note that a divided model with several subcontractors is also possible in a case where both construction and maintenance are included in the contract. In this case, using a number of contractors means that each one is responsible for a set of components of, for example, the road being constructed. For example, one company might be responsible for building and maintaining electronic information systems in a tunnel, while another might be responsible for the road in the tunnel. Even if it is typical for a PPP project to have one (general) contractor, this is not theoretically necessary.

Partnering

As described in Nyström (2005) and Eriksson (2010), partnering can be given a number of more specific interpretations. Their view, and ours, is that partnering should be seen as a way to carry out a certain project in a more collaborative way, opening up for adjustments during the project. This means that partnering should not be seen as a specific procurement contract type, but as something that can be implemented in any type of procurement contract.

The Complete Proposed Framework

Figure 4 shows all the different dimensions of our proposed framework. The idea is that a specific procurement contract could be seen as a specific point on this diagram. Notice that, for the second and third steps in the diagram, there are choices along continuums. The continuum in the second step is between the client being responsible for the entire design; and the contractor being responsible for the entire design, with the client only formulating general demands. The continuum in the third step is between there being one or many contractors.

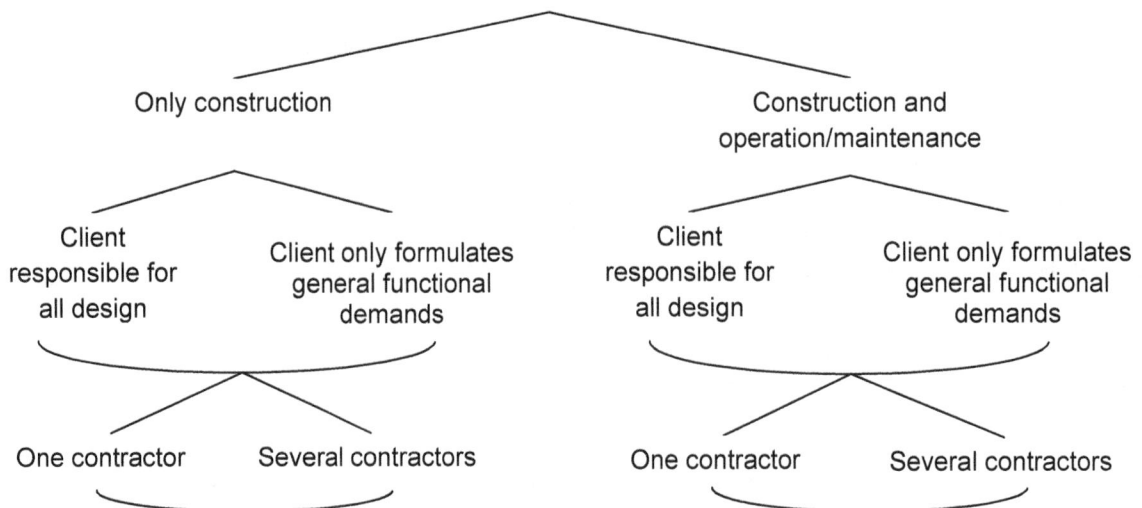

Figure 4: The complete proposed framework (Source: authors)

The Proposed Basic Framework: Payment Method

We propose that payment methods be divided into two main categories depending on whether there is any project-specific risk sharing or not. It can be argued that bonuses in relation to completion time and higher quality can be introduced into all payment methods; therefore, we discuss bonuses separately. Here, we only include payments where the public client pays the contractor. We will not discuss the risk of the client not paying according to the contract, since that can happen in any type of contract. For a PPP project, payment methods are also possible in which the user pays for the use of a road/rail; but we will not discuss these methods here. However, we include a discussion of some special issues that occur in contracts in which construction and maintenance is integrated.

No Project-Specific Risk Sharing

Here there are at least three subgroups:

1. Fixed price: All risk is borne by the contractor/contractors.

2. Fixed price with general indexing (e.g. consumer price index, or a general construction price index): All risk except changes in the general price level is borne by the contractor.

3. Cost-plus contracts: All risk is borne by the client.

Project-Specific Risk Sharing

The cost of a project depends on prices for various inputs, and on how much of each different factor of production is needed (see Brunes & Lind 2014). Especially in more complex projects, there are uncertainties in both dimensions; and there are various ways of sharing these risks.

- Unit price contracts: In this case, the prices for different types of work are fixed in the contract. For at least some dimensions, there are 'variable quantities', which means that the contractor is paid according to the actual quantities, but using the agreed price per unit. This means that the risk related to the prices is born by the contractor (unless prices are correlated with a general index), while the risk related to the quantities needed is borne by the client.

- Sharing in cost increases/cost reductions (Cost sharing): In this case, the basic idea is that there is an 'agreed price', but if the actual cost is higher than this price, a certain share of the increased cost is paid by the client. If the actual cost is below the agreed price, then the price paid by the client is reduced by a certain share of this saving. This means that both price and quantity risks are shared.

Bundled Contracts with Construction and Operation/Maintenance: A Special Issue

These bundled contracts typically cover a 10-30 year period, opening up at least one extra issue concerning the payment method. Should the client pay only a yearly fee - covering both capital costs and operating costs? Or should the client pay a combination of an 'up-front' payment and a yearly fee - the up-front payment when the premise is ready for use (which can be seen as covering the investment cost), and the yearly fee for operation and maintenance? This second model was, for example, used in one of the first road projects of this type in Sweden (Norrortsleden). There appear to be at least two arguments for the second model. The first argument is that this payment method fits better into the government budget system, if the government has a traditional investment budget. The second argument is that if the financing cost for the government is lower, then the total cost will be reduced if a certain sum is paid by the government when construction is finished.

Bonuses and Penalties in Relation to Time and Quality

All contracts specify what the contractor should deliver. In all contracts, it is possible to add bonuses and penalties if the contractor delivers something that is better or worse than what was agreed on in the contract. The simplest versions of bonuses and penalties relate to project completion. However, other measurable indicators can also be used as a basis for bonuses and penalties related to quality. In some cases, it may be more rational to make the contractor pay a penalty rather than having to redo work to reach the desired quality; however, we will not discuss this option further here. Bonuses and penalties for construction contracts may also relate to the guarantee period. These and other bonuses and penalties may also relate to contracted quality, both during the contract period and, for integrated contracts, at the end of the contract period. How the contract handles bonuses and penalties is a complex issue that requires special study.

Payment Methods in Relation to Selected Literature

Our proposed framework covers the different payment methods presented in the literature review that we outlined earlier. What we call *Fixed price* in our framework covers what Gould and Joyce (2011) call *Single fixed price*; what Winch (2010) calls *Fixed price*; and what Ritz and Levy (2013) call *Lump sum*.

Gould and Joyce (2011) further mention *Unit price* contracts and *Cost plus a fee* contracts which are included above. The *Fee based* contract in Winch (2010) is similar to the *Cost plus* contract, and his *incentive* contract covers both what we refer to as *Cost sharing* and the various bonus systems mentioned above. Ritz and Levy (2013) further mention *Cost-plus* and *Bonus* contracts, which are covered by our categories above.

Reflections on the Choice of Contract Type and Payment Mechanism

Potential Problems

From a theoretical perspective, it is clear that all models contain potential problems:

- Integrating construction and maintenance has potential advantages in terms of creating incentives for minimising life-cycle costs. On the other hand, integrating construction and maintenance tends to reduce competition; and it is always difficult to write long-term contracts (e.g. Lind & Borg 2010).

- Leaving the detailed design to the contractor opens up new solutions, and makes it easier to adapt the design to the skills of the contractor. On the other hand, giving the contractor an increased degree of freedom also increases the risk of moral hazard. In addition, it increases the chance of solutions that minimise the short-term cost of the contractor (see Nyström, Lind and Nilsson 2014; Borg 2011). This creates a problem because a risk-averse contractor tends to choose traditional established solutions that might reduce the rate of innovation (Borg 2011).

- Using a general contractor reduces the transaction costs and the risk for the client. On the other hand, this model also reduces competition; and transferring risk to another party always comes with a cost.

In the same way, all payment methods contain potential difficulties:

- Putting all the risk on the contractor can reduce competition, and can be costly if the contractors are risk-averse.

- If a price index is used, it leads to the issue of choosing the right index. In some cases, indexing can create new uncertainties, because the effect of the index is unsure; for example, when different prices change in different directions.

- A cost-plus model that puts all the risk on the client reduces incentives for the contractor.

- Unit price contracts can open up strategic bidding, if contractors think that the actual quantities will differ from the quantities listed in the procurement documents (e.g. Mandell & Nyström 2013; Skitmore & Cattell 2013).

- Sharing divergences from an agreed price makes it necessary to measure the actual cost, which can be difficult. Risk sharing contracts also reduce incentives for the contractor.

- Bonuses that are related to certain specific parameters (e.g. completion on time) may lead to reduced quality in other dimensions, as the contractor focuses on dimensions that lead to a bonus (Milgrom & Roberts 1992, Ch. 7).

How deterministic are the relations?

Given the complexities discussed above, a reasonable research strategy would be to find relationships of the following type: 'In situation X, contract type Y is the best' (e.g. Eriksson & Hane 2014). Warsame, Borg and Lind (2013) however, question whether finding such relationships is really possible. How a certain model works in a specific situation depends on the skill and experience of the parties involved. If an actor believes in 'model A' and is aware of the potential problems in this model, it might be possible for that actor to take measures to control these problems and therefore get good results from 'model A'. A different actor who believes in 'model B' might instead make that model work well in the same situation. A client who has had problems with one model might choose to change to another just to get a new start.

Conclusion

The main contribution of this paper to the existing body of research is our proposed framework for classifying contract types and payment methods. This has advantages in at least three different contexts.

The first advantage of a logical classification system is that it helps the client to make the right decisions. Here are some examples of how our frameworks can be helpful for the decision maker:

- Instead of initially think in terms of a number of contract types (DBB, DB, and PPP) and a choice between them, the framework points out that the first step is to decide what is to be procured. Should the contract only include construction, or should it include both construction and operation/maintenance?

- The next step is to think in terms of: 'Who will do the detailed design of the premise?' Our framework makes it clear that this is not an either/or decision, but rather continuums of alternatives, and that it is rational for a client to regulate certain things in detail, while leaving other things to the contractor.

- The final step in our framework is then to analyse the optimal structure of the contractor side. How many contractors will the client use? Should it be one general contractor or should the responsibilities be divided in one way or the other?

- The framework concerning payment methods also starts from what we believe are the fundamental issues: How should various risks be divided? How should incentives for good behaviour be created? Instead of starting with a long list of different payment methods, the proposed framework pushes the decision maker to start by thinking about the underlying basic issues concerning risk and incentives.

The second context where our framework can be useful is for structuring scientific investigations. Nyström, Lind and Nilsson (2014) present results from a number of evaluations of the effects of using DB contracts instead of DBB contracts. The conclusion is that no pattern can be found. Given our framework, that is not surprising as there is no clear line between DB contract and DBB contracts. When there in reality is a continuum of alternatives for allocating the task of making the detailed design, it might as well be the case that the line is drawn in different ways in different organisations or for different project. This means that what in one case is called a DBB contract may be similar to what in another case is called a DB contract. Then it is not surprising that no significant difference in outcome can be found. In order to make interesting evaluations, it is necessary to go into details of the projects compared, to really find projects where there were large differences between how the design was made, and then compare the effects of these differences.

Finally our framework can be useful for policy formulation, or at least a help to avoid some problematic ways of formulating policies. In recent years, one goal for the Swedish Transport Administration (STA) has been to increase the rate of innovation in infrastructure projects. This has in turn led to a measurable goal that the share of DB projects should be increased at a certain rate. The framework above indicates that this is not a good way to formulate a goal that is to be used to evaluate how successful the administration has been. As, according to our framework, there is no clear line between DBB projects and DB projects, a risk with formulating a goal in this way is that the STA simply re-labels their contracts without making any real changes in the contracts. Our framework can be used to formulate more precise goals, for example in terms of more projects where construction and operation/maintenance is integrated, and more projects where specific parts of the detailed design are left to the contractor/contractors.

References

Bejrum, H. & Grennberg, T. 2003, 'En väg till fungerande hus: Funktionsentreprenader, livscykelekonomi och BOT [A way towards well-functioning houses: Performance contract, life-cycle economy and BOT]', Rapport no. 19, KTHs Bostadsprojekt, Department of Real Estate and Construction Management, KTH Royal Institute of Technology, Stockholm, Sweden. [in Swedish]

Borg, L. 2011, 'Incentives and choice of construction techniques', Licentiate Thesis, Building & Real Estate Economics, Department of Real Estate and Construction Management, KTH Royal Institute of Technology, Stockholm, Sweden.

Brunes, F. & Lind, H. 2014, 'Explaining cost overruns in infrastructural projects: A new framework with applications to Sweden', Working Paper Series, WP201401, Department of Real Estate and Construction Management & Center for Banking and Finance (CEFIN), KTH Royal Institute of Technology, Stockholm, Sweden.

Eriksson, P.E. 2010, 'Partnering: What is it, when should it be used, and how should it be implemented?', Construction Management and Economics, 28 (9), 905-17. doi: http://dx.doi.org/10.1080/01446190903536422

Eriksson, P.E. & Hane, J. 2014, 'Entreprenadupphandlingar - Hur kan byggherrar främja effektivitet och innovation genom lämpliga upphandlingstrategier? [Construction Contracts - How can developers promote efficiency and innovation through appropriate procurement strategies?]', Uppdragsforskningsrapport 2014 (4), Swedish Competition Authority, Stockholm, Sweden. [in Swedish]

Eriksson, P.E. & Laan, A. 2007, 'Procurement effects on trust and control in client-contractor relationships', Engineering, Construction and Architectural Management, 4 (4), 387-99. doi: http://dx.doi.org/10.1108/09699980710760694

FHWA. n.d., 'Project Delivery Methods', Handout - General guidance, US Department of Transportation, Federal Highway Administration, USA.

Gould, F. & Joyce, N. 2011, Construction Project Management, Third Edition, Pearson Prentice Hall, New Jersey, USA.

Leiringer, R. 2003, 'Public-Private Partnerships - Conditions for Innovation and Project Success', Atkin, B., Borgbrant, J. & Josephson, P-E., (eds), Construction Process Improvement, Blackwell Science Ltd., Oxford, 154-65.

Lind, H. & Borg, L. 2010, 'Service-led construction: Is it really the future?', Construction Management and Economics, 28 (11), 1145-53. doi: http://dx.doi.org/10.1080/01446193.2010.529452

Mandell, S. & Nilsson, J-E. 2010, 'A comparison of unit price and fixed price contracts for infrastructure construction projects', Working Papers Series 2010 (13), Swedish National Road & Transport Research Institute (VTI), Sweden.

Mandell, S. & Nyström, J. 2013, 'Too much balance in unbalanced bidding', Studies in Microeconomics, 1 (1), 23-25. doi: http://dx.doi.org/10.1177/2321022213488845

Mattsson, H-Å. & Lind, H. 2009, 'Experience from procurement of integrated bridge maintenance in Sweden', European Journal of Transport and Infrastructure Research, 9 (2), 143-63.

Milgrom, P. & Roberts, J. 1992, Economics, Organization & Management, Prentice-Hall Inc., New Jersey, USA.

Ng, S.T. & Wong, Y.M.W. 2007, 'Payment and audit mechanisms for non private-founded PPP-based infrastructure maintenance projects', Construction Management and Economics, 25 (9), 915-23.

Nyström, J. 2005, 'The definition of partnering as a Wittgenstein family-resemblance concept', Construction Management and Economics, 23 (5), 473-81. doi: http://dx.doi.org/10.1080/01446190500040026

Nyström, J., Lind, H. & Nilsson, P-E. 2014, 'Degrees of freedom in road construction', CTS Working Paper 2014:20, Center for Transport Studies, Stockholm, Sweden.

Ritz, G.J. & Levy, S.M. 2013, Total Construction Project Management, Second Edition, McGraw-Hill Education, NY, USA.

Skitmore, M. & Cattell, D. 2013, 'On being balanced in an unbalanced world', Journal of the Operational Research Society, 64 (1), 138-46.

SOU 2009:24, 'De statliga beställarfunktionerna och anläggningsmarknaden [The governmental purchasing functions and construction market]', Available online: http://www.sweden.gov.se/content/1/c6/12/14/50/1a7ba334.pdf accessed 29 April 2011. [in Swedish]

Warsame, A., Borg, L. & Lind, H. 2013, 'How can clients improve the quality of transport infrastructure projects? The role and knowledge management and incentives', The Scientific World Journal, 2013, Article ID 709423, 8 pages. doi: http://dx.doi.org/10.1155/2013/709423

Winch, G.M. 2010. Managing Construction Projects: An information processing approach, Second Edition, Blackwell Publishing Ltd., UK.

6

Post-Disaster C&D Waste Management: The Case of COWAM Project in Sri Lanka

Gayani Karunasena (University of Moratuwa, Sri Lanka)
Raufdeen Rameezdeen (University of South Australia, Australia)
Dilanthi Amaratunga (University of Salford, United Kingdom)

Abstract

Waste management is considered to be the weakest phase in responding to a disaster. This became apparent when Sri Lanka suffered enormously from the Indian Ocean tsunami of 2004. The City of Galle located on the south coast was severely affected by this event, causing some 4000 deaths and destroying over 15000 houses. The Construction Waste Management (COWAM) project funded by the European Union from 2005-2009 looked at the most sustainable ways of dealing with Construction & Demolition (C&D) waste after a disaster and devised a pilot C&D recycling plant (COWAM Centre) in Galle. This paper reflects on the C&D waste management practices followed by the city authorities during the recovery and reconstruction phase right up to the operation of the COWAM Centre with the intention of seeking best practices for the future. As part of the COWAM case study, semi-structured interviews were conducted with municipal authorities and voluntary organizations to identify the C&D waste management process followed during recovery and reconstruction. Empirical data was collected from actual demolition sites located in Galle to establish the quantity of C&D waste, composition, hazardous substances found, and collection efficiency. Findings revealed that waste was disposed initially into temporary dumping sites and later re-cycled through the COWAM Centre. However, this study found many issues that could have been avoided if Galle Municipal Council had planned and implemented a quick C&D waste management strategy. Key issues which arose were lack of heavy vehicles, lack of manpower, inability to forecast the amount and composition of waste, and inability to identify suitable temporary dumping sites. The characteristics of C&D waste gave a baseline for the design of COWAM Centre. The paper presents a viable approach to overcome issues pertaining to C&D waste management during the aftermath of a disaster through the lessons learned from the COWAM project.

Keywords: Disaster waste, C&D waste, recycling, recovery, reconstruction, Sri Lanka, COWAM Project

Introduction

Disasters create enormous amounts of Construction & Demolition (C&D) waste through destruction of buildings and infrastructure (EPA, 2008; FEMA, 2007). For example the Great East Japan earthquake and tsunami of 2011, Haiti earthquake of 2010, hurricane Katrina of 2005 and Indian Ocean tsunami of 2004 are events that generated large volumes of waste and overwhelmed the existing solid waste management capacities of the local authorities concerned (Basnayake et al., 2005; Luther, 2008; Brown et al., 2010; Shibata et al., 2012). Brown et al. (2011) assert that disaster debris impacts not only on the environment but also rescue and emergency services, provision of lifeline support and socio-economic recovery of the affected areas. Thus, management of wastes created by disasters has become an increasingly important issue that needs to be addressed when responding to a disaster (Thummarukudy, 2012).

Sri Lanka, an Indian Ocean island with 20 million people, experienced a waste management crisis at the aftermath of the Indian Ocean tsunami of 2004. Approximately US$ 5-6 million has been spent in managing tsunami debris in Sri Lanka (Basnayake et al, 2005; UNEP, 2005). Costs to public health and the environment due to prolonged exposure to waste could increase the above estimate significantly (Srinivas and Nakagawa, 2008). Environmentally sensitive coastal dumping sites, waste burning, insufficient landfill capacity, impact on ground water, and lack of coordination were some of the problems encountered by local authorities in the affected areas (Basnayake et al., 2005; Selvendran and Mulvey 2005; Pilapitiya et al., 2006; Srinivas and Nakagawa, 2008).

Galle, the third largest Sri Lankan city and located on the south coast was severely affected by the tsunami with some 4000 deaths and destruction of more than 15000 houses. The Construction Waste Management (COWAM) project funded by the European Union from 2005-2009 examined at the most sustainable way of dealing with the C&D waste generated during the disaster in Galle. It devised a C&D recycling plant called the 'COWAM Centre' to recycle most of these waste residues. This paper reflects on the C&D waste management practices followed by the Galle Municipal Council during the recovery and reconstruction phase right up to the operation of the COWAM Centre with the intention of seeking best practices for the future.

Post-Disaster Waste Management

Management of disaster waste is much harder than ordinary C&D waste as the former are very often mixed and contaminated (Rafee et al., 2008; Kobayashi, 1995). According to the United States' Environment Protection Agency, disaster waste comprises soil and sediments, building rubble, vegetation, personal effects, hazardous materials, mixed domestic and clinical wastes and, human and animal remains (EPA, 2008). It presents a risk to human health from biological, chemical and physical sources (EPA, 2008). The composition of waste differs from disaster to disaster. However, C&D waste is a common type found in every disaster along with automobiles, furniture, vegetative debris, mixed metals, ash and charred wood (EPA, 1995). Baycan and Patterson (2002) classified C&D wastes arising out of a disaster as: recyclable materials (concrete, masonry, wood, metal, and soil), non-recyclable materials (household inventory, organic materials, and other inert materials) and hazardous waste (asbestos, chemicals). According to Kourmpanis et al. (2008), C&D waste is considered to be a priority waste stream and appropriate actions must be taken to make its management effective.

Past disaster experiences show various methodologies have been employed in different contexts. Table 1 illustrates four such situations and the way disaster waste had been managed. It clearly illustrates that the most preferred methodology is recycling even though there are issues with mixed waste, contamination etc. Recycling is the collection and separation of materials from waste and subsequent processing to produce marketable products (Tam and Tam, 2006). Recycling prevents useful material resources being wasted, reduces the consumption of raw materials and reduces energy usage. Numerous materials can be recycled, with the most common being wood, concrete, brick, metals (Ortiz, et al., 2010). Recycling is widely assumed to be environmentally beneficial, although the collection, sorting and processing of materials into new products also entails significant environmental impacts (Klang, et al, 2003). Nonetheless, recycling supplies valuable raw materials to the reconstruction process (Rameezdeen, 2009). In order for a material to be successfully recycled, three major areas need to be taken in to account, namely, the cost-benefit of recycling, compatibility with other materials, and properties of the recycled material (Tam and Tam, 2006).

Baycan (2004) introduced a model for disaster waste management based on the experience of Marmara earthquake of Turkey in 1999. Rubble constituted a major portion of C&D waste. Accordingly, rubble was collected first and transported to temporary dump sites during the recovery period. These materials were transported to recycling plants or landfill sites later depending on the recycling potential. The methodology was criticized as double handling which resulted in high transportation costs. The key principles of this model are;

a.   reduction of quantities of waste for final disposal,

b.   conservation of natural resources, and

c.   minimization of prolonged exposure and resulting problematic environmental impacts.

Similarly, Eerland's (1995) waste management model of the Kobe earthquake gives a clear picture of how segregation was carried out. Initially the waste was segregated using separation plants of the capacity of 50 tonnes per hour. Screening, wind sifting, hand picking and belt separators were some of the technologies used. Finally, the separated materials were sent to recycling and reuse (Eerland, 1995).

Table 2 illustrates the waste management strategies used during recent natural disasters in Sri Lanka. As in many developing countries, the most commonly used strategy was either open dumping or land filling. Uncontrolled dumping of disaster waste can have a significant negative impact on public health and the environment.

Table 1: Disaster waste management examples

Disaster	Amount of waste	Strategies used for management	Issues encountered
Marmara Earthquake Turkey	13 million tons	• Recycling plant • 17 dump sites	• Large quantities of reinforcement bars cause operational problems in the recycling plants • illegal dumping at coastal areas
Kobe Earthquake Japan	15 million tons	• A small proportion recycled • Majority ended up in land reclamation	• Segregation is time-consuming and costly
Beirut, Lebanon	4 million tons	• A stationary recycling plant	• Problems arising due to the unclean nature of disaster debris
Kosovo	10 million tons	• A mobile recycling plant • Decentralized depots for collection and storage	• Spread of the damage over a large rural area

Source: Baycan and Petersen, 2002; Earland, 1995; Pasche and Kelly, 2005

Table 2: Disaster waste management strategies used locally

Disaster	Strategies used for management	Issues encountered
Tsunami, 2004	▪ Local authorities, volunteers and land owners removed debris ▪ Recycling plants were used in Galle and Batticaloa ▪ Incineration ▪ Land filling	▪ Lack of awareness of waste management techniques ▪ Unplanned landfill sites in environmentally sensitive areas ▪ Illegal landfill sites ▪ Lack of capacity to handle large quantities of waste ▪ Inadequate funds for waste management ▪ Lack of coordination
Floods (frequent)	▪ Open burning ▪ Open dumping ▪ Land filling	▪ Illegal dumping on roadsides, vacant land or river banks
Landslides (frequent)	▪ Recycling ▪ Open dumping ▪ Land filling	▪ Poor collection methods ▪ Lack of capacity to handle large quantities of waste

Source: Pilapitiya et al., 2006

Methodology

The aim of this study was to investigate the C&D waste management practices that followed the relief, recovery, rehabilitation and reconstruction phases of a disaster in order to learn lessons for the future. Case study method proved to be the most appropriate, as it provides access to the real-life context of disaster waste generation, collection and handling (Yin, 2003). It provides a rich data set based on experiences and explanations of the people and organizations involved. As it begins with a theoretical framework described above, it has the ability to test the existing theories or concepts. However, the major limitation of the present study is that it cannot be based on multiple cases as disaster situations are very rare. The study is based on a single disaster, the 2004 Indian Ocean tsunami. The City of Galle was selected as the case study site due to the accessibility of information and the fact that a Construction Waste Management (COWAM) project was implemented during 2005-2009. COWAM was funded through the EU-Asia Pro Eco II B – Post Tsunami Programme and was led by the TuTech Innovations GmbH of Hamburg, Germany. The project partners in Sri Lanka were the Galle Municipal Council (GMC) and the University of Moratuwa. As part of the case study, interviews and C&D data collection on actual demolition sites were carried out using a retrospective study design described below (Robson, 2011).

First, interviews were conducted among GMC officials and volunteer organizations who elicited information on the waste management strategies and the challenges they encountered during the post-disaster phase. The interviewees were the Lordship of Mayor of Galle, Deputy Mayor, 4 elected members, Municipal Commissioner, Municipal Engineer,

Medical Officer, 3 technical officers belonging to the waste management division, 2 Public Health Inspectors, and one officer from each of the 5 volunteer (non-government) organizations involved in relief and recovery operations. The duration of these 19 interviews varied from 30 minutes to one hour, conducted in the Sinhalase language, and recorded with the permission of the interviewee. The sample is a reasonably good representation of the persons involved in waste management operations having both officials and volunteers. The limitation of the interviews is that the opinion of the general public could not be obtained due to the very large sample frame and the resource constraints of the study. Interviewees were asked to describe each phase of the waste management process based on a semi-structured guideline that reflected the theoretical base of the study. Any personal bias was removed through triangulation as a result of more than one representation from different groups (as an example, the politicians' version was not only that of the ruling party but also from the opposition; 3 out of the 4 elected members were from the opposition). Key themes emerging from the interviews were identified and cross-checked among different groups (politicians, officials, volunteers). Code-based content analysis technique was used to analyze the data using NVivo (Version 7) software. Coding structure was prepared focusing mainly on two themes as given in Figure 1.

Second, waste benchmarking was carried out in order to establish a baseline for design of the COWAM Centre and as a valuable database for future use. Twelve demolition projects operated by local demolition companies were selected to obtain data to develop a 'waste index' and to assess the composition of C&D waste pertaining to local conditions. This is a quasi prospective-retrospective study design to obtain data from a current situation that could be related to a similar past event (Robson, 2011). Direct observations coupled with actual data from these companies enabled a 'controlled' evaluation of the amount of waste generated in a demolition operation. Direct observation as a data collection technique has been successfully used in similar studies by various researchers (Lu et al., 2011; Gavilan and Bernold, 1994; Poon et al., 2004). A researcher was stationed during the whole demolition operation to record the number of truck loads of waste handled and the volume of each truck in order to obtain the 'waste index' of a building as follows.

$$\text{Waste Index (C)} = \frac{\text{Volume of waste generated in a building (W)}}{\text{Gross Floor area (GFA)}}$$

Where,

W = Volume per truck load (V) x Number of truck loads (N)

45 random waste samples were obtained from these demolition operations into timber boxes measuring 1 m^3 to ascertain its composition and the proportion of hazardous materials present. The quantification of waste in these samples was carried out using source evaluation methodology. The use of source evaluation for waste quantification is very popular among researchers (Lu et al., 2011; Ekanayake and Ofori, 2000; Bossink and Brouwers, 1996). A comparison of the waste collection techniques among these demolition companies was made possible due to some using 'source separation' and others 'commingled collection'. Out of the 12 projects, 8 used source separation and 4 commingled collection. Cost, revenue and profit of each demolition operation were employed to gauge the efficiency of the two methodologies. The cost and revenue of a recycling operation consisted of the following components.

Cost = Cost of demolition + Cost of separation + Cost of recycling + Cost of disposal of the residue

Revenue = Payment from client + Revenue from selling recycled products

Findings from the case study are discussed in the following section.

Picture 1. Sampling Tool

Figure 1: Coding structure used

Findings

Waste generation and collection

Post-disaster recovery at Galle was mainly handled by the GMC. Soon after the disaster, due to its unmanageable volume, the municipality decided to dump all waste wherever space was available. There were no records of types and amounts of waste deposited in these temporary sites. Neither was there a mechanism to forecast the amount of waste generated by the destruction. In order to overcome this issue should future disasters occur, the waste benchmarking exercise described in the methodologies section was carried out as a part of the COWAM project. According to Poon et al. (2004, p.677), a 'waste index' could be used to estimate the quantities of waste generated due to a demolition operation. Using the formula given in methodologies section, waste indices were calculated for the twelve demolition projects; 10 single storey and 2 two-storey residential buildings. Mean waste indices for these two types were found to be 0.34 m^3/m^2 and 0.73 m^3/m^2. Using these values, the volume of C&D waste generated as a result of 15000 houses being destroyed in Galle could be estimated.

Municipal workers and volunteers initially cleared the access routes by dumping waste at the curbside. Subsequently, using municipality trucks these wastes were transported to temporary dumping sites in a very ad hoc manner. The main issues encountered at this stage were: firstly, the identification of dumping sites close to Galle; and secondly, lack of trucks and workmen to collect and transport waste to these identified sites. There was also the technical issue of demolishing partially destroyed buildings as they posed a severe risk to the public. The municipality did not possess equipment to carry out such an operation. Volunteers and builders helped to clear these partially destroyed buildings. During collection, it was not sure whether 'source separation' or 'commingled collection' would be used considering a host of uncertainties surrounding a disaster event. Though some districts like Batticaloa had segregated C&D waste prior to dumping, Galle municipality dumped it without segregation. Financial data collected through the 12 case studies revealed that profits are higher in source separation compared to commingled collection as given in Table 3. However, in a disaster, it is rarely economic efficiency that becomes the deciding factor. Interviews revealed that commingled collection was the most practical strategy to use.

Table 3: Comparison of waste handling methodologies (n=12)

Type of handling	Mean revenue per GFA (in SL Rupees)	Mean cost per GFA (in SL Rupees)	Mean profit per GFA (in SL Rupees)
Source Separation	3132	1087	2045
Commingled Collection	1883	970	913

In order to implement a waste management strategy, it is very important to know the composition of waste being generated. Table 4 illustrates the composition based on the 45 samples of 1m^3 timber boxes used in this study. More than 50% of the waste consists of bricks, cabok (a Laterite brick) and normal clay brick. This poses some challenges to recycling which is described in the next section. It is also important to understand the proportion of hazardous waste in the total waste stream in addition to its composition and the potential impact on human and environmental health. The study found that approximately 14.2% of C&D waste could contain different hazardous materials. As this is very substantial, suitable measures have to be taken in dealing with these wastes. Table 5 gives the composition of these materials that could contain hazardous substances. Asbestos contributed about 19% of hazardous C&D waste. Time and effort was spent on removing hazardous substances at the recycling stage which dramatically reduced the financial viability of recycling operation.

Processing

Waste from the temporary dumping sites was transported to the COWAM Centre for processing during the rehabilitation and reconstruction stages of the disaster. Had the waste been separated, it would have been of much higher quality. In order to sustain the local recycling market, readily re-usable products were sent to the market directly without being taken to the COWAM centre. The study found that there is a small but thriving recycled product market in Galle. Thus, COWAM Centre received mostly the mixed waste that did not have a market value. This measure helped the local recycling market to be revived as a result of the disaster.

Hazardous substances were removed prior to separation and crushing. The COWAM Centre crushed these mix wastes to produce aggregates for use in local roads. The low quality of local roads in Galle requires high levels of maintenance, imposing high costs on the municipal budget (5.3% of annual budget). This is particularly problematic after monsoon seasons when many potholes must be repaired. Apart from the associated labour and monetary costs, regular road maintenance also causes traffic jams. Road construction and maintenance also uses a significant share of natural material resources that can be partially replaced by recycled C&D products. The aim of this measure is to simultaneously reduce the environmental impact of road construction and maintenance while improving road and maintenance quality, reducing maintenance requirements and municipal costs. The residue needs to be sent to permanent landfills after recycling was estimated to be only 4-7% of the waste bought to the COWAM Centre.

Table 4: Composition of C&D waste in Galle (n=45)

Material	Composition (%)
Cabok	29.85
Bricks	28.67
Mortar	15.24
Concrete	6.77
Clay	2.36
Timber	1.57
Asbestos roofing sheets	1.17
Clay roof tiles	0.94
Ceramics	0.82
Plastic	0.50
Wires	0.25
Steel	0.05
Glass	0.01
Mixed waste	11.80
TOTAL	100.00

Table 5: C&D waste that could contain hazardous substances (n=45)

Material	Composition (%)
PVC, uPVC	21.28
Electrical waste	19.75
Asbestos materials	18.68
Paint	15.43
Treated timber	13.26
Gypsum boards	6.23
Waster proofing material	3.09
Sealants, varnish, etc.,	1.83
Roofing cement	0.44
TOTAL	100.00

Capacity constraints of GMC

The case study revealed a few issues with regard to capacity constraints of the GMC which prevented an effective waste management during the post-disaster stage. These observations could be generalized across other local authorities as some of them are due to lack of a national strategy and a framework for C&D waste management in Sri Lanka. These constraints are:

- Lack of a framework or statutory guidelines that could be enforced during a disaster,
- Lack of technical knowhow on C&D waste management,
- Lack of funds, resources, and equipment, and
- Coordination issues.

An in-depth review of the national disaster management policy shows that there is no provision, model, or a framework for disaster waste management (Disaster Management Act No. 13 of 2005). Solid waste is handled by the National Environmental Act of 1981 and C&D waste is considered 'inert'. Its management is not given a high priority compared to the municipal solid waste (Rameezdeen, 2009). The organizational structure of the GMC is arranged in such a way and the priority of top management is placed solely in dealing with the day to day municipal waste. Technical training programs, both foreign and local, which are used to train and maintain an effective labor force, were not sought in the area of C&D waste management. The Technical Officers and the Public Health Inspectors were not interested in C&D waste; they do not envisage a career in C&D waste management; and as a result do not feel training is required in that area. Lack of funds, resources and equipment have prevented GMC in going beyond the basic obligations of the municipal council. Even the municipal waste is collected with great difficulty and the interviewees see no reason why they should go beyond their mandate. Coordination issues were highlighted as a major constraint particularly during disasters. While many NGOs had the manpower and were willing to become involved in tsunami waste disposal, GMC never approached them.

Discussion

Even though a national level agency was established after the tsunami disaster as the peak body for disaster management in Sri Lanka, it is the local authority that has to attend to most of the initial relief and recovery operations. Thus, the capacity of the local authority to deal with disaster waste becomes crucial. Therefore, this study particularly concentrated on local authority level waste management process that followed the tsunami disaster of 2004. The case study on Galle Municipal Council (GMC) found several issues pertaining to the capacity of dealing with C&D waste. The main concern was the neglect of C&D waste in the day-to-day operations of the GMC; lack of interest, knowhow, funds, resources and equipment to deal with it; and serious coordination problems during a disaster.

This case study demonstrated the use of waste benchmarking for planning. The 'waste index' could become a very useful tool in rapid need assessments undertaken immediately after a disaster. According to the International Federation of Red Cross Societies, rapid need assessments are undertaken typically within the first week of a disaster to establish the immediate survival needs of the affected people, which includes waste disposal among a host of other needs such as food, shelter, medical care, safe drinking water, and psychosocial support. If national level disaster organizations such as Disaster Management Centre can provide standard 'waste indices' in their guidelines for different types of buildings and infrastructure, local authorities can easily estimate the amount of C&D waste generated during a disaster. Composition of C&D waste and the extent of hazardous substances present are useful information for planning. As more than 50% of C&D waste emanated from bricks, source separation became very difficult and the recycled output of mix waste was of low quality.

Based on the case study, a process model is suggested for disaster C&D waste management as given in Figure 2. Accordingly, waste collection needs to be handled in two stages; first, a preliminary curb-side collection in order for emergency operations to resume, and second, a planned operation of either source separation or commingled collection. FEMA (2007) supports this observation and recommends the creation of on-site temporary collection centres. Galle's experience shows that source separation, though ideal, is very difficult to organize and handle. In disaster situations, source separation was found to be inefficient and impractical as clean-up and recovery is the priorities rather than recycling (Peterson, 2004). If waste can be source separated, it could then be deposited in temporary mono landfills until the local authority is ready for processing. Commingled collection will lead to mixed waste and the subsequent plant based separation will be costly. The COWAM project increasingly sought the collaboration of local recycling companies to benefit from the operation and therefore only handled mixed waste that was not considered valuable. These mixed wastes were crushed to form aggregates for use in local roads.

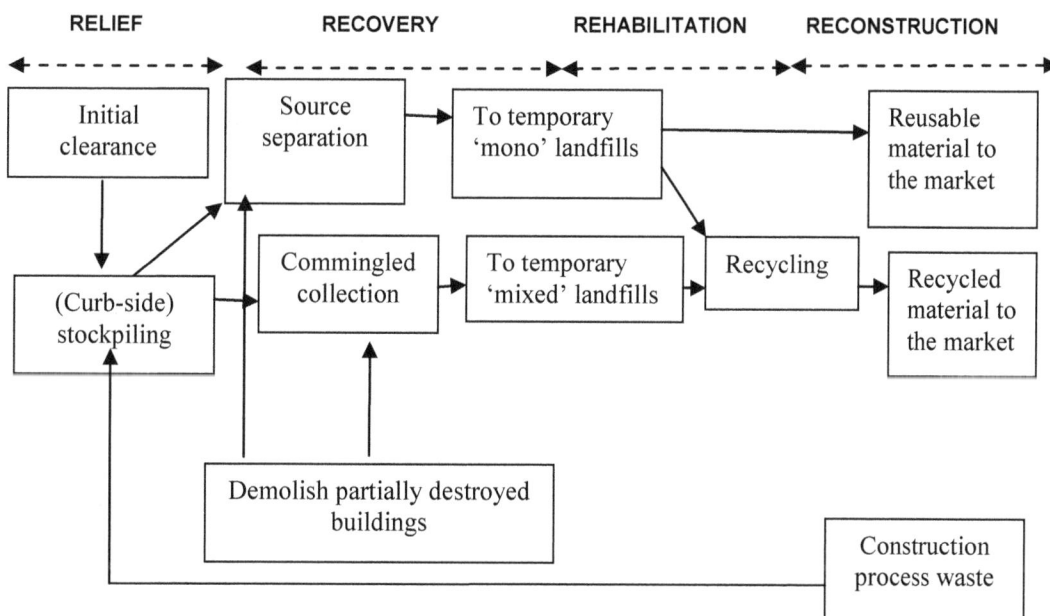

Figure 2: Proposed C&D waste management model

Conclusions

The COWAM case study provides some lessons for future disaster C&D waste management at the local authority level. The main issues encountered were the lack of capacity to handle an enormous quantity of waste, difficulty in estimating the quantity and composition of waste, difficulty in identifying temporary dumping sites and coordination among different parties involved. In order to overcome these issues, waste benchmarking and publication of those results was suggested. Collection and processing of waste should be done in two stages, with temporary landfills serving as an intermediate solution. Source separation coupled with mono-landfills will provide a high quality output when the final recycling is carried out.

However, limited resources may force a local authority to carry out commingled collection with mixed landfills. Encouraging the participation of local recycling companies will boost the local recycling market. A process model encompassing the above stages summarized the major findings.

References

Basnayake, B.F.A., Chiemchaisri, C., Mowjood, M.I.M. 2005, Solid wastes arise from the Asian tsunami disaster and their rehabilitation activities: case study of affected coastal belts in Sri Lanka and Thailand, Tenth International Waste Management and Landfill Symposium, Sardinia.

Baycan, F. and Petersen, M. 2002, Disaster waste management-C&D waste, in: ISWA, ed. Annual conference of the international solid waste association, 8-12 July 2002 Istanbul. Turkey:ISWA.

Baycan, F. (2004), Emergency planning for disaster waste: A proposal based on the experience of the marmara earthquake in Turkey., [Online]. Available at http://www.corporate.coventry.ac.uk [Accessed on 21 August 2012].

Bossink, B.A.G. and Brouwers, H.J.H. 1996, Construction waste: Quantification and source evaluation, Construction Engineering and Management, 122(1), 55-60.

Brown, C., Mike, M., Seville, E. 2010, Waste management as a lifeline?, A New Zealand case study analysis, International Journal of Disaster Resilience in the Built Environment 1(2): 192-206

Brown, C., Mike, M., Seville, E. 2011, Disaster Waste management: A review of articles, Waste Management 31: 1085-1098.

Eerland, D.W. 1995, Experience with the construction and demolition waste recycling in the Netherlands – Its application to earthquake waste recycling in Kobe. In: IETC, ed. International symposium on earthquake waste, 12-13 June Osaka. Shiga: UNEP, 72-85.

Ekanayake, L.L. and Ofori, G. 2000, Construction material waste source evaluation, Proceedings of Strategies for a Sustainable Built Environment, Pretoria.

Environmental Protection Agency (EPA). 1995, Characterization of building related construction and demolition waste in the United States. EPA 530-R-98-010, 1998.

Environmental Protection Agency (EPA). 2008, Planning for Natural Disaster Waste, [Online] available at: http://www.epa.gov/CDmaterials/pubs/pndd.pdf [Accessed 10 June 2012].

Federal Emergency Management Agency (FEMA). 2007, Public Assistance: Waste Management Guide, [Online] available at: http://www.fema.gov/government/grant/pa/demagdes.html [Accessed 10 June 2012].

Gavilan, R.M. and Bernold, L.E. 1994, Source evaluation of solid waste in building construction, Construction Engineering and Management, 120(3), 536-552.

Klang, A., Vikman, P.and Brattebo, H. 2003, Sustainable management of demolition waste an integrated model for the evaluation of environmental, economic and social aspects, Resources, Conservation and Recycling 38, 317-334.

Kobayashi, Y. 1995, Disasters and the problems of wastes. In: IETC, ed. International symposium on earthquake waste, 12-13 June 1995 Osaka. Shiga: UNEP, 6-13.

Kourmpanis, B., Papadopoulos, A., Moustakas, K., Stylianou, M., Haralambous, K.J., and Loizidou, M. 2008, Preliminary study for the management of construction and demolition waste. Waste management and research, 26(3), 267-275.

Lu, W., Yuan, H., Li, J., Hao, J.J.L., Mi, X., and Ding Z. 2011, An empirical investigation of construction and demolition waste generation rates in Shenzhen city, South China, Waste Management, 31, 680-687.

Luther, L. 2008, Managing Disaster Waste: Overview of Regulatory Requirements, Agency Roles, and Selected Challenges, Congressional Research Service [Online] Available at: http://wikileaks.org/wiki/CRS-RL34576 [Accessed 25 February 2012].

Ministry of Environment and Natural Resources (MENR). 2005, Post Tsunami Environmental Assessment in Sri Lanka: Recommendations for Environmental Recovery, Ministry of Environment and Natural Resources, Colombo, Sri Lanka.

Pasche, A. & Kelly, C. 2005, Concept Summary: Disposal of Tsunami generated waste, UNDAC/Sri Lanka.

Petersen, M. 2004, Restoring waste management following disasters. In: IF, ed. International conference on post disaster reconstruction, 22-23 April UK. Coventry: IF Research group.

Pilapitiya, S., Vidanaarachchi, C.,Yuen, S. 2006, Effects of the tsunami on waste management in Sri Lanka, Waste Management, 26(2), pp. 107–109.

Poon, C.S., Yu, A.T.W., Wong, S.W. and Cheung, E. 2004, Management of construction waste in public housing projects in Hong Kong, Construction Management and Economics, 22(8), 675-689.

Rafee, N., Karbassi, A.R., Nouri, J., and Safari, E. 2008, Strategic management of municipal debris aftermath of an earthquake. International Journal of Environmental Research, 2(2), 205-214.

Rameezdeen, R. 2009, Construction waste management: Current status and challenges in Sri Lanka, COWAM publication, Colombo.

Robson, C. 2011, Real world research: A resource for users of social research in applied settings, 3rd Edition, John Wiley, United Kingdom.

Selvendran, P.G. and Mulvey, C. 2005, Reducing solid waste and groundwater contamination after the tsunami, Daily news, Tuesday, 15 February, 2005.

Shibata,T., Solo-Gabriele, H., Hata, T. 2012, Disaster waste characteristics and radiation distribution as a result of the Great East Japan Earthquake, Environmental Science & Technology, 46, 3618–3624.

Srinivas, H. & Nakagawa, Y. 2008, Environmental implications for disaster preparedness: Lessons Learnt from the Indian Ocean Tsunami, Journal of Environmental Management, 89(1), pp 4-13.

Tam, V.W.Y. and Tam C.M. 2006, Evaluation of existing waste recycling methods: A Hong Kong study, Building and Environment, 41, 1649-1660.

Thummarukudy, M. 2012, Disaster waste Management: An overview, in Shaw, R. and Tran, P. (eds.), Environment disaster linkages, Emerald Group Publishing limited,

United Nations Development Programme (UNDP). 2005, Tsunami Recovery Waste Management Programme (TRWMP) NAD-Nias, UNDP, Indonesia.

Yin, R.K. 2003, Case research design: design and methods. 3rd ed., London: Sage publications.

7

An Examination of the Structure of Sustainable Facilities Planning Scale for User Satisfaction in Nigerian Universities

Abayomi Ibiyemi, Yasmin Mohd Adnan, Md Nasir Daud (University of Malaya, Malaysia)

Martins Adenipekun (Lagos State Polytechnic, Nigeria)

Abstract

Universities are under increasing pressure to demonstrate that continuous performance improvement is being delivered for user satisfaction, but the importance of facilities planning as a student-staff focused tool needs to be emphasised. This research sought answers to questions relating to the underlying structure of sustainable facilities planning and user satisfaction, and the number of factors that make up the facilities planning scale. Three universities from the south-western part of Nigeria were selected randomly using ownership structure to define the cases: University of Lagos, Akoka, Lagos, Ladoke Akintola University of Technology, Ogbomoso and Joseph Ayo Babalola University, Ikeji Arakeji, each representing the Federal, State, and Private ownership. A questionnaire survey was used on a random sample of 651 staff and students from the three universities. Six hundred questionnaires were retrieved (response rate of 92.2%). An exploratory factor analysis was used to understand the responses and the interrelationships. The results showed a two-factor solution of 'locational advantages and user needs' and 'adequacy of facilities/functional connection and four core determinants for acceptance. It is concluded that universities should factor student-staff focus points into their facilities planning schemes to optimise their service deliveries. The study contributes to the discussion on factor structure of sustainable facilities planning scale with a focus on students and staff of universities.

Keywords: Facilities planning, universities, data structure, factors, Nigeria.

Introduction

The concept of facilities planning evolved in response to the need for a more dynamic planning process, but, in the education sector, it poses a number of challenges not common in other sectors. Many challenges confront the effectiveness and efficient operation of educational facilities in the university system in recent times. Resultantly, several calls have been made to give more attention to the management of university facilities for improved qualitative outputs and user satisfaction.

Facilities in many universities in developing countries are becoming obsolete, and grossly inadequate to achieve the objectives of those universities. Some universities operate on temporary campuses that bear no resemblance to contemporary university expectations. Consequently university social environments do not appear to synchronise with expected student/teacher/staff relationships. There have been incessant strikes, lock outs and student agitations which sometimes resulted in destruction of support facilities. The management of universities' real estate assets has become increasingly inefficient, because many institutions do not have facilities planning units where the concept of facilities planning (FP) can be utilised in an operational capacity. Given the financial and resource constraints under which the universities must manage, it is essential that students and staff expectations are understood and measured (Robathan 1996; Adewunmi, Omirin & Koleoso 2012). The basic thrust of sustainable facilities planning (SFP) was described by Van Mell (2005) as the provision of precise building or buildings needed to support strategic goals for satisfying corporate objectives and user satisfaction. Steiss (2005) emphasized that a systematic planning effort is vital to decision-making about construction and financing of strategic facilities and that values of sustainable facilities planning SFP should be demonstrated.

The management of university facilities transcends the problem of personnel, janitorial, transportation, or mere sanitation issues, but includes maximisation of efficiency in time savings, space, capital budgeting, staff welfare, teamwork and improvement in general productivity (Shayler 2010). The common appearance of decay, neglect, under-utilisation, over-utilisation and abandonment of structures is an indication of sparse strategic FP at the design and construction stages of university facilities. Van Mell (2005) stated that to achieve university corporate objectives, a detailed planning of every facet of university facilities is desirable. Literature abounds with evidence to justify the importance of the integration and application of FP to the management of facilities and user satisfaction (Marmolejo 2007; Fareo and Ojo 2013), but there is also the need to justify the importance of FP as a student and staff-focused tool for continuous performance improvement of universities. User satisfaction is a reflection of the degree to which users feel that their education environment is helping them to achieve their goals. However, some university management can be neglectful at times in considering student and staff viewpoints. This is a valid research problem that necessitated the development of the SFP scale. The overall competitive advantage index of an education system could diminish with having obsolete or inappropriate facilities if the importance of SFP is not articulated for proper understanding and implementation.

Study Objective and Significance

The study seeks to answer the following two questions:

- What is the underlying factor structure of sustainable facilities planning (SFP) and user satisfaction?

- What is the importance of SFP as a student and staff-focused tool for continuous performance improvement of universities?

This study is significant because it is of immediate relevance to national educational interests in developing countries, and helps to provide the basic index for further analytical studies in facilities planning in university settings.

Study Context

This study is demonstrated in the context of south-west (SW) Nigeria. The choice of Nigeria is informed by the common appearance of decay, neglect, under-utilisation and over-utilisation of facilities on many of their university campuses, which prima facie indicate lack of strategic and sustainable facilities planning. The SW zone consists of Ogun, Oyo, Ondo, Osun, Lagos and Ekiti states. These states share similar educational, socio-cultural, economic and political characteristics, and they constitute the fastest growing region in education when compared with the geo-political zones in the country. There are thirty-one approved universities within the South-West zone (NUC 2011). University of Lagos (UNILAG) is located at Akoka, Yaba, while the Medical Campus of the College of medicine is located a few kilometers from the main campus at Idi-Araba, Surulere, on the mainland of Lagos. The university, established in 1962, has residential, office and academic facilities and services for both staff and students. It has fourteen academic units comprising a broad range of professional faculties and schools. Most faculties are located on the main campus. Ladoke Akintola University (LAUTECH) was established in 1990. The main campus is at Ogbomosho in Oyo state. The campus is the site of the university's administration, as well as home to five faculties and the post-graduate school. The other campus is located in Osogbo, home to the College of Health Sciences, and Faculties of Medicine and Surgery, Medical Laboratory

Technology and Nursing. Joseph Ayo Babalola University (JABU) is a private Nigerian university located in Ikeji-Arakeji in Osun state. It was established by the Christ Apostolic Church (CAC) Worldwide in 2002. The university is a fully residential institution, with about seven faculties. There are 38,000 registered students in UNILAG, 25,000 in LAUTECH and 15,000 in JABU. Staff populations are 4,000, 3,000, and 2,800 respectively (NUC 2011).

Review of Literature

Facilities management (FM) is complex in scope (Paxman 2007). It integrates the people, process and the place. FM embodies inter-related job responsibilities that include long-range facility planning, annual tactical planning, facility financial forecasting and management, real estate acquisition and disposal, interior space planning, work specifications, installation and space management, architectural and engineering planning and design, new construction and/or renovation works, maintenance and operations maintenance of the physical plant, telecommunication integration, security and general administrative services such as food services, records management, reprographics, transportation and mail services, health, safety and out sourcing (Rondeau, Brown & Lapides 1995). It anchors and integrates all of the job responsibilities together to design a corporate policy objective. It is therefore seen as an emerging field that incorporates many interacting terminologies (Van Mell 2005). FM poses a strong relationship with other disciplines, such as space planning, architecture, interior design, environmental psychology, real estate, systems engineering, human resource management, information systems management, project management, and building service engineering (Adegoke and Adegoke 2013). Central to basic FM functions and activity areas is planning and programming. Every aspect of facilities management requires detailed planning to achieve its objectives. Long (5 years) and short (1 year) term programs should be aligned with corporate and departmental operating plans and these should incorporate major activities (Paxman 2007; Krizek et al. 2012). Minimisation of facilities costs, tracking and pattern of changes also exert certain impact on firms' goals, underscoring planning as an imperative in all activities of FM (Somovoa 2007). ASBO (2003) asserted that effective management starts with planning that synthesises collaborative interfaces, but cautioned that planning can result in real problems when large capital investment sums are misappropriated. Bennett (2010) summarized effective facilities planning and management of public sector property as speculative and one which requires a conceptual framework to operate. He therefore investigated the validity of facilities planning by considering the relationship between customer and business led strategies and how different strategies affect facilities planning. A prototype strategic model for future use decision making process where it is necessary to consider facilities needs and challenges in business was proposed. Bennett restated that facilities or assets strategies within this context are centered on business strategies and organizational aspirations rather than making current assets dictate direction of strategic business goals. The business strategies and organizational approach allows the definition of a destination whereas the current assets approach provides an indication of the starting point and consequently the length of journey. The business and organizational focus rather than facilities or asset focus allows strategies to be developed which, once in place, can provide an informed platform for the more organized facilities management model to come into play with cost-in-use planning, maintenance planning, space planning (Shayler 2010; Wright & Olesand 2007; Burud 2010), budget planning, asset planning, operation planning, systems and software planning (Dingley 2008), capital planning, campus planning (Koppelman 1975), contract facility planning, environmental planning and scenario planning (Sekula 2010; IJFM 2010) and organization planning (Gladwell 2000; Lepkova & Uselis 2013).

The American institute of Architects (2011) and UNM (2014) concurred that a sustainable facility plan should answer the following questions about an organization's real estate portfolio: How much space will the organization need to carry out its activities - *quantity*? What kinds of space will it need - *type*? During what time period will the organization need it - *timing*? How will it procure it, buy, build, or lease - *portfolio mix and duration*? What will it cost - *budget*? Where does the space need to be located - *location*? Which groups need to be located near each other - *affinity and allocation*? How will the organization deal with unplanned changes in demand for space - *hedging and exit strategy*? What mechanisms will it employ to let users see the actual costs of their occupancy; forecast their future needs accurately, honour their promises concerning occupancy, and use space efficiently? (*Internal business model, including items such as internal leases and transfer charges*) How will the facilities contribute to the core business of the organization through their effect on marketing, employee recruiting, and employee retention - *corporate identity, location, and amenities*? Can the organization reduce total real estate costs per person - *density, design standards, and alternative office spacing*? Can it affect employee productivity and rate of production throughput by the design of the facilities - *Effectiveness and productivity*? In their view, FP process for capital projects for campus development should consist of (1) Identifying, budgeting, documenting, justifying, and prioritizing built environment needs, in which facilities plan identifies the type, quantity and location of spaces needed by the department or college through an in-depth analysis of existing facilities; (2) Understanding of program growth, projected student volumes, research aspirations; and developing an achievable and affordable plan to meet the college's current and future facility needs, with an outcome of achieving academic initiatives and supporting the mission and goals of the Institutions. Figure 1 illustrates the SFP process.

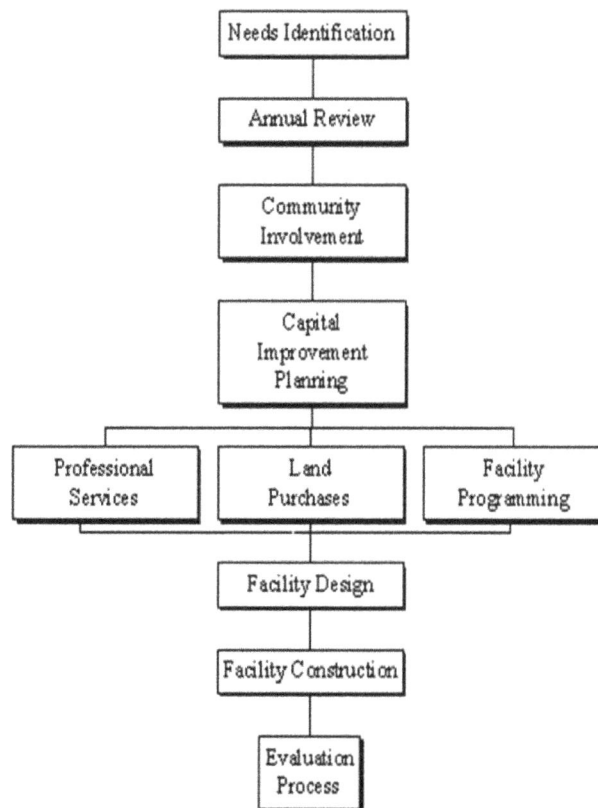

Figure 1: The SFP Process (Adapted from AIA, 2011 and UNM, 2014)

UNM remarked that a full needs assessment takes account of all the relevant academic programs, research and/or class listings specific to the targeted area of concern, assessment of existing space/building conditions, utilisation of current spaces, and projected volumes as validated from the Provost office. This is the review stage in which, according to IFMA (2009), the Facilities Manager, Planner and designers begin to identify the gaps in current facility needs with long-term needs. Issues that include workforce demographics, manufacturing processes, organizational structure and culture, community and government regulatory requirements, market positioning, and capacity rates and volumes are applied to balance the gaps. The immediate community may be involved to facilitate stakeholders' contributions and cooperation. Scenario planning is conducted by selecting several external forces and imagining related changes that might influence the organization, such as the global marketplace, technology, change in regulations, and demography. For each potential change, the best, worst, and reasonable cases are reviewed in order to suggest what the organization might do, or potential strategies, in each of the three scenarios to respond to each change. The facility design and evaluation, which the design of the facilities is based on, is a re-evaluation of the strategies. To deploy the use of computer to aid FP in this sector would no doubt involve preparation of strategic integration of user requests. Room use between the teaching, off teaching hours, leisure-based activities, use of multi-purpose halls, squash and tennis courts, bars and other spaces used by students, staff or outside organizations will differ widely. Some organizations hire out facilities to outside organization during vacations to increase revenues, particularly accommodation and meeting spaces. These require adequate planning of every aspect of the facilities to integrate their various strategic foci for corporate objectives to be achieved. In the literature, it is assumed that the application of FP (and FM in general) is focused on supporting primary processes and contributing to achieving organisational goals (Atkin & Brooks 2000; Barret & Baldry 2003) the physical setting of which can aid or hinder the accomplishment of internal organizational goals, and user satisfaction (Bitner 1992). A clear expression of this is the large number of FM-related studies that have been conducted focussing on different aspects of its added value for primary processes (Williams 1996; Krumm, Dewulf & De Jonge 1998; Amaratunga & Baldry 2000; Salonen 2004; Wauters 2005; Lindholm & Leva¨inen 2006; De Toni et al. 2007; Chotipanich & Nutt 2008; De Vries, De Jonge & Van Der Voordt 2008; Adewunmi, Omirin & Koleoso 2012), especially quality (e.g. customer satisfaction), users satisfaction regarding time (e.g. response time), risk (e.g. safety, reputation) and relationship quality (e.g. alignment). It has already been established that FP value concerns a trade-off by the users between benefits, costs and risks. In this study, the scope of FP focuses on the contribution to user satisfaction, and its added value can be defined as the user perceived contribution of the different facility services to satisfying their needs.

Main knowledge gaps in FM literature relating to higher institutions include studies on the students perspective of housing facilities for continuous improvement in student housing satisfaction (Sawyer & Yusof 2013), and assessing the magnitude of deterioration in African tertiary institutions and their relative lack of conduciveness to learning (Adegoke and Adegoke 2013). Literature has also not provided evidence of the factor levels of contributions of SFP scale to user satisfaction, nor justified the importance of the FP as a student and staff-focused tool for continuous performance improvement of SW Nigeria Universities. This study is an investigation that seeks to address that gap.

Research Method

The research adopted a survey approach. The survey research offered the scope for large representative sampling of students and staff from where reliable information can be extracted about user satisfaction from the two heterogeneous populations. This is preferred to interviews and observations, which have very limited scope in this regard. SW Nigeria has the highest concentration of universities (31) in the country and a good mixture of Federal (6), State (9) and Private (16) universities (JAMB 2010). Three of the universities were randomly selected from each cluster of ownership and control structure – one from federal, state, and private ownership respectively.

The secondary information used for this study was gathered from internet sources, relevant journals, seminar and conference papers, and monographs. Others include materials from text books, base maps, and government publications. The National Universities Commission (NUC), the physical development offices of the respective universities, works and services, and other relevant departments were observed directly for triangulation. The primary data collection used questionnaires. A total number of 651 students and staff were served with close-ended questionnaires (217 questionnaires per university). 600 questionnaires were retrieved (92.2% response rate). Respondents were selected using the random technique and categorized into" Students" and "Staff". The variables of study were identified in current literature (American institute of Architects, 2011 and UNM, 2014) and classified into three factors for measuring user satisfaction as shown in Table 1.

Table1: Classification of factors for measuring user satisfaction

Main Variables	Sub-Variables/References	Code	Type
Location advantages and user needs	• Location advantage of facilities and other services in meeting user needs.	LAF	Dummy
	• Demand for the use of facilities by staff and students for socio-economic functions.		
Adequacy of facilities and functional connection	Repairing condition of facilities	RCF	Dummy
	• Maintenance Adequacy of Facilities.	MAF	Dummy
	• Use of facilities plans in the management of facilities.	UFP	Dummy
	• Functional connection of facilities to one another with respect to user needs.	FCF	Dummy
	• Functional design relationship of facilities to one another.	FDR	Dummy
	• Adequacy of facilities and other services in meeting user needs	AFF	Dummy
	Effect of facilities condition on user performance.	EFC	Dummy
Response time	• Response time to repair dysfunctional facilities.	RTDF	Dummy
	• Impact of dysfunctional facilities on operation of other facilities connected to it.	IMDF	Dummy
	• Impact of facilities condition on work efficiency.	IMFWE	Dummy

33.9%, 34.3% and 31.8% of students' respondents are from UNILAG, LAUTECH, and JABU respectively, while 31%, 34.5%, and 34.5% of staff respondents are from UNILAG, LAUTECH, and JABU respectively. The distribution of questionnaires among students and staff is fair and representative. No university has less than 30% of the total number of questionnaires retrieved.

The campus facilities, units and services considered in this study include: The main library, departments' libraries, division of students affairs, university health services, sports centres, the bursary, registry, internal audit, consults, and ventures, bookshops and press. Others are car park, guest houses, conference center, central industrial liaison placement, main auditorium, botanical/ zoological gardens, community pharmacy, academic planning unit, guidance and counselling unit, estate units, media/corporate affairs, works and physical planning unit, hydraulic research unit, alumni relations unit, legal unit, security unit, records and quality assurance, students' halls of residence, and the senate building complex. Others are lecture theatres, shopping complexes, sports complex, senate building, SUG building, and banks. These are the composites of what constitute real estate whether tangible or intangible, services and other benefits real or abstract in form that contribute to the achievement of the corporate objectives of the institutions.

Methods of analysis: Frequency distribution of field data and the exploratory factor analysis-principal component analysis were used. The reliability of the three subscales was assessed using Cronbach alpha for internal consistencies: All the three factors (a) to (c) have coefficients ≥.7 on the average (Devellis 2003; Kline 2005; Pallant 2011).

Results and Analyses

Each questionnaire was designed in close-ended form to enable respondents to provide responses to 12 independent variables that constituted FSP scale. The SFP scale was employed in the assessment of user satisfaction, the outcome variable. The analysis of the questionnaire and the response rates are shown in Table 2.

Table 2: Frequency and Mean Scores of Questionnaire Responses from the Universities (Analysis of Responses, 2014)

Scale	LAF	DUF	RCF	MAF	UFP	FCF	FDR	AFF	EFC	RTDF	IMDF	INFWE
	Locational Advantages & User needs			Adequacy of facilities and functional connection						Response time		
Students Respondents												
Very Poor (1)	87	72	90	68	75	98	107	95	114	92	84	9
Poor (2)	70	56	49	73	61	50	49	55	50	46	56	51
Fair (3)	17	31	32	39	51	25	22	28	25	26	25	34
Good (4)	60	60	52	87	57	51	54	59	37	58	73	55
Very Good (5)	96	111	107	63	86	108	98	93	104	108	92	101
Total	330	330	330	330	330	330	330	330	330	330	330	330
Mean score	3.1	3.3	3.1	3.0	3.1	3.1	3.3	3.0	2.9	3.1	3.1	2.9
Staff Respondents												
Very Poor (1)	69	72	89	59	69	85	78	69	76	87	79	89
Poor (2)	54	36	31	57	56	38	40	56	55	35	56	48
Fair (3)	29	21	26	35	30	17	27	30	17	14	30	27
Good (4)	46	41	40	59	56	38	45	56	36	29	41	39
Very Good (5)	72	100	84	60	59	92	80	59	86	105	64	67
Total	270	270	270	270	270	270	270	270	270	270	270	270
Mean Score	3.0	3.2	3.1	3.0	2.9	3.0	3.2	3.0	3.0	3.1	2.8	2.8

A total of 330 students responded to all the variables of study, while 220 staff respondents were recorded across the universities.   All the samples were subjected to EFA. The Likert 5 points summated rating scales was used and it ranged from Very Poor (1) to Very Good (5). For the purpose of interpreting Table 2, the mean scores were calculated and related to the scale labels, very poor (1) to very good (5) based on ∑(scale x frequency)/number of respondents.

Respondents were asked to assess the location benefits/advantage of facilities to users. Mean scores of 3.0 and 3.1 obtained for student and staff respondents indicated that location benefits/advantages of facilities with respect to needs of users were generally fair. The locations of facilities were fairly suited to the needs of users.

They were also asked to assess the demand for the use of facilities in the university for social and economic functions. Mean scores of 3.3 and 3.2 obtained for student and staff respondents indicated fair public demand for the use of the various universities' facilities. It can be interpreted that there is demand for use of facilities across the universities.

The study sought information on the state of repair conditions of facilities in each of the selected universities. Mean scores of 3.1 from the respondents indicated fair state of repair conditions. This study also investigated the adequacy of maintenance of facilities in the universities. Mean scores of 3.0 were obtained, indicating that maintenance of facilities is fairly adequate.

We asked respondents to assess their respective universities on the use of maintenance plans in the management of universities facilities. Mean scores of 3.1 and 2.9 were obtained for student and staff respondents respectively. It indicated fair use of facilities plan in their various universities from the students' perspective, while it is poor from the staff perspective. However, generally, the use is not optimal.

We also asked respondents to assess the functional connection of facilities to one another for smooth and unhindered operation in their respective universities. Mean scores of 3.1 and 3.0 obtained for student and staff respondents show that the functional connection of facilities to one another with respect to the needs of users is fair.

The research also elicited information on the functional design relationship between facilities to each other with respect to the efficient discharge of users' works. Mean scores of 3.3 and 3.2 obtained for student and staff respondents indicated fair functional design relationship of facilities to one another for the purpose of allowing users to discharge their duties effectively and with comfort.

We also implored respondents to assess whether the facilities were adequate in quantity and services for the work users expect those facilities to perform. Mean scores of 3.0 were obtained, indicating that maintenance of facilities is fairly adequate. All the universities generally had fairly adequate facilities to meet needs of users.

With respect to weighing the effect of condition of facilities on users' performance, rating were allocated for the universities on: the period of time it takes to respond to request to repair dysfunctional facilities; assessing the impact of broken-down and out of operation facilities on the operation of other facilities on performance in the work environment; and the impact of facilities condition on their work efficiency. A 43.8% rating was achieved. Mean scores of 3.1; 3.1 and 2.8; 2.9 and 2.8 were obtained respectively. While period of time it takes to respond to request to repair dysfunctional was minimal, the effect of condition of facilities on users' performance, and facilities and the impact of facilities condition on their work efficiency was generally poor.

Exploratory Factor Analysis (EFA)

The EFA is based on correlation matrix; that all the variables are correlated to some degree. Table 3 is the basis of the application of the EFA-Principal Components. EFA has been used in this study to understand and identify the pattern of responses of students and staff of three universities completing closed-ended questionnaires. The items measuring similar things can be identified, and therefore forms the structure of replies to the questionnaire. A coherent set of data which addresses the research questions is achieved. Studies with similar themes that have also adopted the EFA-Principal Components are those of Pallant and Bailey (2005) which used it for the assessment of the hospital anxiety and depression scale in musculoskeletal patients; Zemering (2009) in ascertaining the perception of government workers to sustainability programs in the US; Addae-Dappah, Hiang & Shi (2009), in the assessment of perception of investors and users to sustainable property features in Singapore; and Oven and Pekdemir (2006), in establishing office rent determinants in Istanbul.

PART I - Oblimin Rotation of four-factor solution (Default in SPSS22)

The procedure suppresses the presentation of any factor loadings with values less than .3 (Pallant 2011). The correlation matrix of the SFP sub-variables is shown in Table 3 where there are correlation coefficients greater than .03. Specifically, it appears that there are nine groups of variables that are strongly intercorrelated, i.e. having coefficients greater than .500: (i) FDR/AFF (.666) (ii) UFP/MAF (.651) (iii) RCF/MAF (.612); (iv) FCF/MAF (.612); (v) RTDF/UFP (.558) (vi) AFF/EFC (.542); (vii) FDR/EFC (.533) (viii) RCF/UFP (.528); and (ix) UFP/FCF (.517).   The largest correlations occur in (i) and (ii).   The Kaiser-Meyer-Olkin (KMO) value is greater than .6 (KMO=.860) (Kaiser 1970; Addae-Dappah, Liow Kim Hiang & Neo Yen Shi. 2009), and the Bartlett test of sphericity is significant at p ≤ .05 (p=.001) (Bartlett 1954). The indication is that factor analysis is appropriate (Robert Ho 2006; Field 2009; Pallant 2011; Howitt and Crammer 2011).   Table 4 shows the Total Variance Explained result.

Table 3: SPPSS22 (DEFAULT) Correlation Matrix of SFP sub-variables

		LAF	DUF	RCF	MAF	UFP	FCF	FDR	AFF	EFC	RTDF	IMDF	INFWE
Correlation	LAF	1.000	.426	.238	.300	.207	.159	.278	.242	.244	.138	.231	.199
	DUF	.426	1.000	.203	.278	.235	.213	.265	.217	.179	.112	.188	.126
	RCF	.238	.203	1.000	.612	.523	.465	.307	.378	.298	.326	.277	.191
	MAF	.300	.278	.612	1.000	.651	.612	.423	.466	.378	.420	.337	.285
	UFP	.207	.235	.523	.651	1.000	.517	.336	.356	.285	.558	.259	.160
	FCF	.159	.213	.465	.612	.517	1.000	.300	.343	.265	.352	.293	.182
	FDR	.278	.265	.307	.423	.336	.300	1.000	.666	.533	.201	.371	.275
	AFF	.242	.217	.378	.466	.356	.343	.666	1.000	.542	.277	.382	.269
	EFC	.244	.179	.298	.378	.285	.265	.533	.542	1.000	.234	.243	.161
	RTDF	.138	.112	.326	.420	.558	.352	.201	.277	.234	1.000	.140	.159
	IMDF	.231	.188	.277	.337	.259	.293	.371	.382	.243	.140	1.000	.422
	INFWE	.199	.126	.191	.285	.160	.182	.275	.269	.161	.159	.422	1.000

As in Table 4, we considered Kaiser's criterion in determining how many factors to extract: Factors or components that have Eigenvalues greater than 1 (Kaiser's criterion) are reported. The first four components recorded Eigenvalues ≥ 1 (4.594, 1.400, 1.111, and 1.028). These four components explain a total of 67.78% of the variance (38.29%, 11.67%, 9.26%, and 8.56% respectively). Components having Eigenvalues less than 1 are ignored because such factors consist of uninterpretable error variation (Pallant 2011; Howitt and Crammer 2011). The Scree Plot, when examined, has a break in the size of Eigenvalues for the factors occurring after the second factor. The curve is also fairly flat after the second factor. The indication is that the scree plot supports a two-factor solution. However, the other factors were analyzed further as their Eigenvalues were also greater than 1. Table 5 shows the unrotated loadings of each of the items of the four components/factors.

As in Table 5, the four components are then obliguely rotated and the loadings of the 12 variables on these 4 factors are as shown in the table. It can be seen that most of the items loaded quite strongly, i.e. above .4 (Pallant 2011; Howitt and Crammer 2011) on two components, LAF (.456 and .609), and DUF (.426 and .684). We then used the criterion values randomly generated from similar sized data sets and obtained from Monte Carlo Parallel Analysis, to take a decision on the number of factors: If the size of Eigenvalue was greater than the criterion value, the factor was retained, and if less, the factor was rejected (Pallant 2011). The summary of PCA and Monte Carlo parallel analysis for decision making is shown in Table 6.

Table 4: Total Variance Explained

Component	Initial Eigenvalues			Extraction Sums of Squared Loadings			Rotation Sums of Squared Loadings[a]
	Total	% of Variance	Cumulative %	Total	% of Variance	Cumulative %	Total
1	4.594	38.287	38.287	4.594	38.287	38.287	3.742
2	1.400	11.665	49.952	1.400	11.665	49.952	3.214
3	1.111	9.259	59.212	1.111	9.259	59.212	2.090
4	1.028	8.564	67.776	1.028	8.564	67.776	2.162
5	.743	6.190	73.966				
6	.613	5.107	79.074				
7	.549	4.578	83.652				
8	.511	4.257	87.909				
9	.468	3.904	91.813				
10	.390	3.253	95.066				
11	.316	2.635	97.701				
12	.276	2.299	100.000				

Extraction Method: Principal Component Analysis.

a. When components are correlated, sums of squared loadings cannot be added to obtain a total variance.

Table 5: Component Matrix

	Component			
	1	2	3	4
MAF	.815			
UFP	.721	-.456		
AFF	.717		-.319	
FDR	.685	.399		
RCF	.681			
FCF	.665	-.351		
EFC	.601			-.422
IMDF	.545	.329		.500
RTDF	.537	-.464		
DUF	**.426**		**.684**	
LAF	**.456**	.328	**.609**	
INFWE	.428	.321		.677

Extraction Method: Principal Component Analysis.

Table 6: Summary of PCA and parallel analysis for decision making

Component No	Actual Eigenvalue from PCA	Criterion value from MC Parallel analysis	Decision
1	4.594	1.2389	Accept
2	1.400	1.1778	Accept
3	1.111	1.268	Reject
4	1.028	1.0851	Reject
5	.743	1.0496	Reject

The result of parallel analysis supported our decision from the scree plot to retain only two factors for further investigation. Further to making a final decision concerning the number of factors, we examined the Pattern Matrix table, Table 7. The table shows the items loading on the 4-factor solution with 5 items loading above .3 on component 1, 3 items on component 2, 2 items on component 3, only one item on component 4. Ideally, we would have liked 3 or more items loading on each component. This further supports only two factors. Using the SPSS22 default option, we obtained a four-factor solution, hence, we needed to go back and 'force' a two factor solution. This is contained in Part II.

Table 7: Pattern Matrix

	Component			
	1	2	3	4
UFP	.862			
RTDF	.758			
FCF	.736			
MAF	.732			
RCF	.681			
EFC		.872		
FDR		.819		
AFF		.802		
DUF			.867	
LAF			.813	
INFWE				.891
IMDF				.747

Extraction Method: Principal Component Analysis.
Rotation Method: Oblimin with Kaiser Normalization.[a]
a. Rotation converged in 6 iterations.

PART II – Oblimin Rotation of two-factor solution (Adjusted in SPSS22)

As in Table 8, the Total Variance Explained – 49.95% of the variance is explained, compared with 67.78% explained by the four-factor solution. After rotating the two-factor solution, the new Component Correlation Matrix, the Pattern, and the Pattern Matrix Tables were re-examined. The strength of the relationship between the two factors is a moderate intercorrelation (r=.464). This gives us information that the correlation between the two components are low, and that we should also expect similar solutions from varimax rotation. We therefore reported oblimin rotation further in Table 9. Table 9 shows the combination presentation of the pattern matrix, structure matrix, communalities tables for PCA with oblimin rotation of two-factor solution of SFP items. ±0.30 is the minimum level of practical significance. Values less than ±0.30 indicate that the item does not fit well with the other items in the component (Pallant 2011; Tabachnick & Fidell 2007).

Table 8: Total variance explained for the 12 sub-variables

Component	Initial Eigenvalues			Extraction Sums of Squared Loadings			Rotation Sums of Squared Loadings[a]
	Total	% of Variance	Cumulative %	Total	% of Variance	Cumulative %	Total
1	4.594	38.287	38.287	4.594	38.287	38.287	3.731
2	1.400	11.665	49.952	1.400	11.665	49.952	3.744
3	1.111	9.259	59.212				
4	1.028	8.564	67.776				
5	.743	6.190	73.966				
6	.613	5.107	79.074				
7	.549	4.578	83.652				
8	.511	4.257	87.909				
9	.468	3.904	91.813				
10	.390	3.253	95.066				
11	.316	2.635	97.701				
12	.276	2.299	100.000				

Extraction Method: Principal Component Analysis.
a. When components are correlated, sums of squared loadings cannot be added to obtain a total variance.

Table 9: Pattern matrix, structure matrix, communalities tables for PCA with oblimin rotation of two-factor solution of the 12SFP items.

Item	Pattern Coefficients		Structure Coefficients		Communalities
	FACTOR 1 Component 1	FACTOR2 Component 1	FACTOR 1 Component 1	FACTOR2 Component 1	
FDR	**.785**	-.017	**.793**	-.381	.629
AFF	**.704**	-.136	**.767**	-.462	.602
IMDF	**.636**	-.003	**.658**	-.371	.406
EFC	**.619**	-.084	**.637**	-.297	.439
LAF	**.582**	.049	**.560**	-.221	.315
INFWE	**.559**	.058	**.532**	-.201	.286
DUF	**.481**	-.018	**.489**	-.241	.239
UFP	-.021	**.862**	.379	**.853**	.728
RTDF	-.136	**.703**	.561	**.834**	.504
MAF	-.222	**.731**	.385	**.751**	.733
FCF	.048	**.728**	.431	**.733**	.565
RCF	.117	**.679**	.218	**.700**	.548

Note: Bolded items indicate major loadings for each item.

The pattern coefficients show the factor loadings of each of the variables. The main loadings on Component 1 are FDR, AFF, IMDF, EFC, LAF, IMFWE and DUF. The items on Component 2 are UFP, RTDF, MAF, FCF and RCF. Communalities gives information about how much of the variance in each item is explained. IMFWE and DUF have .286 and .239 respectively on communalities. These two items have values that are less than ±0.30, and also show the lowest loadings on Component 1 (.556 and .481). We may therefore use this information to remove the two items from the scale in order to increase the total variance explained, should a confirmatory factor analysis be contemplated. MAF (.733), UFP (.728), FDR (.629), and AFF (.602) have the highest coefficients.

Summary of Findings

The 12 items of the SFP scale were subjected to EFA analysis. Prior to performing EFA, the suitability of data for factor analysis was assessed. Inspection of the correlation matrix revealed the presence of many coefficients of .3 and above. The Kaiser-Meyer-Olkin value was .86, exceeding the recommended value of .6 (Kaiser 1970; 1974) and Bartlett's Test of Sphericity (Bartlett 1954) reached statistical significance, supporting the factorability of the correlation matrix. Principal components analysis revealed the presence of four components with eigenvalues exceeding 1, explaining 38.28%, 11.67%, 9.26% and 8.56% of the variance respectively. An inspection of the screeplot revealed a break after the second component. Using Catell's (1966) scree test, it was decided to retain two components (components 1 and 2) for further investigation. This was further supported by the results of parallel analysis, which showed only two components with eigenvalues exceeding the corresponding criterion values for a randomly generated data matrix of the same size (12 variables × 600 respondents). The two-component solution explained a total of 49.95% of the variance, with Component 1 contributing 31.25% and Component 2 contributing 17.0%. To aid in the interpretation of these two components, oblimin rotation was performed. The rotated solution revealed the presence of simple structure, with both components showing a number of strong loadings and all variables loading substantially on the two components. There was a weak negative correlation between the two factors (r = .464). The results of this analysis support the use of MAF, UFP, FDR, and AFF and these can be regarded as core determinants of SFP for user satisfaction in SW Nigeria Universities.

Findings from descriptive analysis were that locations of facilities were fairly suited to needs of users, and there is demand for use of facilities across the universities. Mean scores ranged from 2.8 to 3.3. Repairing condition is fair, but maintenance of facilities is not adequate. There is poor use of facilities planning in the management of their facilities, although functional connection of facilities to one another with respect to the needs of users is fair. Generally, all the universities have fairly adequate facilities to meet the needs of users. While the period of time it takes to respond to request to repair dysfunctional facilities needs to be optimised, the effect of condition of facilities on users' performance, and also that of facilities condition on their work efficiency are poor. These findings correlated moderately with the researchers' direct observations.

Discussion

The result of this study reveals high levels of components 1 (locational advantage and user needs) and 2 (facilities adequacy and functional connection) in the assessment of user satisfaction. The two-factor solution obtained in this study meets the expectations of the researcher by contributing an investigation into the factor structure of SFP which the researcher has been unable to find in extant and recent literature. The result suggests that modifications to the original structure of SFP are necessary when using the scale in a sample of students and staff of universities. Item 5 (DUF) should not be included in the calculation of the SFP scale because it has the lowest loadings of pattern and structure coefficients of .481 and .489 respectively (see Table 9). The adequacy of facilities and functional connection sub-scale remains consistent and can be used. It is evident that university management cannot afford to be negligent in considering student and staff viewpoints in their facilities planning operations, as most decisions have direct implications for user comfort and satisfaction, and can also be a catalyst for performance improvement of the universities. Empirically, the result justifies the importance of FP as a user-focused tool. However, it has been argued that the respondents might not have adequate knowledge about the overall facilities within the universities. The possible response to this is that the students and staff are informed respondents, who despite possibly having limited access to records relating to facilities, budgeting, and space allocation issues, can nevertheless perceive relationships between these facilities, and how the resource use and allocation have enhanced or diminished their comfort and satisfaction. Their perception, when recorded on the appropriate scale, provides a reliable data set summary which can be triangulated for internal validity. If SFP is well designed and implemented in a stakeholder-integrated fashion that optimises resource use, the corresponding positive effects will resonate even at the lowest levels of management. The study agrees with Steiss (2005) that contribution to the discussion on factor structure of sustainable facilities planning scale should be demonstrated and documented.

Conclusion

It is desirable that strategic and sustainable facilities planning in universities factor student-staff focus points into facilities maintenance management, costs in use, space and general operations planning through an open feedback mechanism. If it does not, optimum university service delivery for organisational effectiveness cannot be expected. Structure details could be generalised while specific details could vary from campus to campus. This implies continual clarification of the structure characteristics for documentation in a manner that engenders a reliable information base. The aim and purpose of the study, to specify the structure of sustainable facilities planning scale in relation to student-staff user satisfaction by identifying the underlying factors and variables inter-relationships, has been achieved. The research questions have been answered. The two underlying factors have been determined, and the interrelationships among the variables explained. The study compares favourably with Steiss (2005), but goes further to propose the student-staff dimension as its main contribution to knowledge and practice. The utilisation of factor analysis as an effective technique to elicit dominant factors in a scale is in concordance with extant and recent literature which enabled the scales and sub-scales to be examined. Importantly, the need to understand the underlying factors underscores the relevance of sustainable facilities planning to user satisfaction, and could provide a platform to operationalise sustainable facilities planning practices for organisational effectiveness. This paper concludes that a well-structured study could also provide empirical information as a life belt for best practices in management planning, physical plant planning, financial planning, total institutional plan, and evaluation of facilities program, response timing, and annual facilities review. The levels of Factors 1 and 2 detected and accepted suggest that they are dominant factors in the sustainable facilities planning scale. However, researchers who are considering using it in a sample of students and staff of universities should flexibly adjust the sub-scale to the climes of the university under their study. EFA supported the presence of the two sub-scales, but suggests that item 5 be removed. The new direction of thinking about the research problem should be to integrate student-staff viewpoints into universities' facility plans as a corollary to community involvement, and the optimisation of facilities use through structured re-evaluation of design and strategies. The respondents were constrained by limited access to classified information, such as facilities acquisition plans and costs. Although, such limitations may impinge on scope for generalisation, they are ineffectual in altering the reliability and validity of the findings. Further research will be necessary to establish cut-off points for a revised one item factor 1 by triangulating with a structured interview.

References

Addae-Dapaah, K, Hiang, L.K & Shi, N. Y. 2009, 'Sustainability of sustainable real property development', Journal of Sustainable Real Estate, 1 (1), 203-225.

Adegoke, B.F. & Adegoke, O.J.T. 2013, 'The use of facilities management in tertiary institutions in Osun State, Nigeria', Journal of Facilities Management 11 (2), 183-192.

Adewunmi, Y, Omirin, M. & Koleoso, H. 2012, 'Developing a sustainable approach to corporate FM in Nigeria', Facilities 30 (9), 350-373.

Amaratunga, D. & Baldry, D. 2000, 'Assessment of facilities management performance in higher education properties', Facilities, 18 (7/8), 293-301.

American Institute of Architects 2011, The Architect's handbook of professional practice, USA, AIA.

Association of School Business Officials International, ASBO 2003, 'Planning Guide for maintaining School Facilities', http://www.ed.gov/pubs/edpubs.html, Accessed 2 September 2014.

Atkin, B. & Brooks, A. 2000, Total facilities management, Blackwell, Oxford, UK.

Barret, P.S. & Baldry, D. 2003, Facilities management: Towards best practice, Blackwell, Oxford, UK.

Bartlett, M.S. 1954, 'A note on the multiplying factors for various chi square approximations', Journal of the Royal Statistical Society, 16 (Series B), 296-298.

Bitner, M.J. 1992, 'Services capes: The impact of physical surroundings on customers and employees', Journal of Marketing, 56 (3), 57-71.

Burud, S. 2010, Working out of the box, Facilities Management Resource, FMLink Online, Retrieved from www.FMJONLINE.com, Accessed 16 August 2014.

Catell, R.B. 1966, 'The scree test for number of factors', Multivariate Behavioral Research, 1 (2), 245-276.

Chotipanich, S. & Nutt, B. 2008, 'Positioning and repositioning FM', Facilities, 26 (9/10), 374-378.

De Toni, A.F, Fornasier, A, Montagner, M. & Nonino, F. 2007, 'A performance measurement system for facility management: The case study of a medical service authority'. International Journal of Productivity and Performance Management, 56, 5/6 417/435.

De Vries, J.C, De Jonge, H. & Van Der Voordt, T.J.M. 2008, 'Impact of real estate interventions on organisational performance', Journal of Corporate Real Estate, 10 (3), 208-223.

Devellis, R.F. 2003, 'Scale development: Theory and applications', Thousand Oaks, California, Sage.

Fareo, D. & Ojo, O. 2013, 'Impact of facilities on academic performance of students with special needs in mainstreamed public schools in Southwestern Nigeria', Journal of Research in Special Educational Needs, 13 (2), 159-167.

Field, A. 2009, 'Discovering statistics using SPSS', Thousand Oaks, California, Sage.

Herman B, Kok, M.P. & Mobach Onno S.W 2011, 'Facilities design', Journal of Facilities Management, 9 (4), 249-265.

Ho, R. 2006, Handbook of univariate and multivariate data analysis and interpretation with SPSS, Taylor & Francis, US.

Howitt, D. & Crammer, D. 2011, Introduction to SPSS statistics in psychology: For versions 19 and earlier, Pearson, Edinburg, UK.

IFMA 2009, The strategic facility planning: A white paper, International Facilities Management Association, UK.

IJFM 2010, A model of workplace environment satisfaction: A survey instrument, FMLink, Facilities management resources, United Kingdom.

JAMB 2010, Joint admissions and matriculations board, eBROCHURE, Joint Admissions Matriculations Board, Abuja, Nigeria.

Kaiser, H. 1970, 'A second generation Little Jiffy', Psychometrika, 35 (4), 401-415.

Kaiser, H. 1974, 'An index of factorial simplicity'. Psychometrika, 39 (1), 31-36.

Kline, T.J.B. 2005, 'Psychological testing: A practical approach to design and evaluation', Thousand Oaks, California, Sage.

Koppelman, J.D.C.L, In Lawal (2000), Ile-Ife, ILCO Books Publishers, 'Urban planning and design criteria', In: Lawal, M.I, (cd), Estate development practice in Nigeria, 1975 ILCO Books, Nigeria.

Krizek, K.J, Newport, D, White, J. & Townsend, A.R. 2012, 'Higher education's sustainability imperative: how to practically respond?' International Journal of Sustainability in Higher Education, 13 (1), 19-33.

Krumm, P.J.M, Dewulf, G. & De Jonge, H. 1998, 'Managing key resources and capabilities: pinpointing the added value of corporate real estate management', Facilities, 16 (12/13), 372-379.

Lepkova, N. & Uselis, R. 2013, 'Development of a quality criteria system for facilities management services in Lithuania', Procedia Engineering, 57, 697-706, DOI: 10.1016/j.proeng.2013.04.088.

Lindholm, A.L. & Levainen, K.I. 2006, 'A framework for identifying and measuring value added by corporate real estate', Journal of Corporate Real Estate, 8 (1), 38-46.

Marmolejo, F. 2007, Higher education facilities: Issues and trends: OECD.

National Universities Commission (NUC) 2011, List of accredited universities in Nigeria: National Universities Commission.

Olesand, N. 2010, The evolving workspace, Facilities Management Resources, Available info@failink.com, Accessed 15 August 2014.

Oven, D. & Pekdemir, A. 2006, 'Office rent determinants utilizing factor analysis', Journal of Real Estate Finance and Economics, 33, 51-73, doi: http://dx.doi.org/ 10.1007/s11146-006-8274-5.

Pallant, J. 2011, SPSS survival manual, Allen & Unwin, NSW.

Pallant, J. & Bailey, C. 2005, 'Assessment of the structure of the hospital anxiety and depression scale in musculoskeletal patients', Health and Quality of Life Outcomes, 3 (82).

Paxman, D. 2007, Facilities Management in practice, IFPI Ltd. Kent, UK.

Robathan, P. 1996, Intelligent building performance facility management: Theory and practice, Spon Press, London.

Rondeau, E.P, Brown, R.K. & Lapides, P.D. 1995, Facility Management, John Wiley & Sons NY.

Salonen, A. 2004, 'Managing outsourced support services: observations from case study', Facilities, 22 (11/12), 317-322.

Sawyer, P.T. & Yusof, N.A. 2013, 'Students satisfaction with hostel facilities in Nigerian Polytechnics', Journal of Facilities Management, 11 (4), 306-322.

Sekula, M. 2010, 'Strategic facility planning: Now more important than ever', Facilities management resources, Facilities Management Journal

Shayler, S. 2010, 'Calculating for change', Facilities management resource

Somorova, V. 2007, 'The task of the facility management in real estate development', Vadyba/Management m, 16-17 (3/4).

Steiss, A. 2005, Strategic facilities planning: Capital budgeting and debt administration, Google books, UK.

Tabachnick, B.G. & Fidell, L.S. 2007, Using multivariate statistics, Pearson Education, Boston, USA.

UNM 2014, University of New Mexico planning and campus development.

Van Mell, D. 2005, What is facility plan? Available: http://www.vanmell.com/Articles/whatisafacilityplan.pdf, Accessed on 2 September 2014.

Vossler, C.A. & Kerkvliet, J. 1999, 'A criterion validity test of the contingent valuation method: Comparing hypothetical and actual voting behaviors for a public referendum', Journal of Environmental Economics and Management 45 (2003), 631-649.

Wauters, B. 2005, 'The added value of facilities management: benchmarking work processes', Facilities, 23 (3/4) 142-151.

Williams, B. 1996, 'Cost-effectiveness facilities management: A practical approach', Facilities, 14 (5/6), 26-38.

Wright, R. & Olesand, N. 2007, Paradigm shifts in property management and work, Work place trends: Managing New Work Styles, London.

Zeemering, E. 2009, 'What does sustainability mean to city officials?' Urban Affairs Review, 45, 247-273, doi: http://dx.doi.org/.1177/1078087409337297.

Dimensions of Organisational Culture in Quantity Surveying Firms in Nigeria

8

Ayokunle Olubunmi Olanipekun, Joseph Ojo Abiola-Falemu & Isaac Olaniyi Aje

Federal University of Technology, Akure, Nigeria

Abstract

The functionalist paradigm of organisational culture (OC) views culture as a variable subject to conscious manipulation and control in order to solve organisational challenges. Therefore, this paper provides information on how OC is a solution to the challenges in Quantity Surveying firms (QSFs). This was achieved by eliciting the dimensions of OC in forty two QSFs in Lagos, Nigeria, which are the business, people and external environment dimensions. The paper concludes that OC is a relevant solution to the identity and management related challenges in QSFs. Specifically, the paper informs on the implications of business and people dimensions of OC as a solution to the identity challenges, as well as on the implication of the external environment dimension of OC to the management challenges. Based on the findings, recommendations are directed at the management and employees QSs in QSFs and Quantity Surveying researchers.

Keywords: Dimensions of Organisational culture, Quantity Surveying firms, functionalist paradigm, identity, management

INTRODUCTION

Quantity Surveying Firms (QSFs) are service based that manage finance related issues for clients in the construction industry (Abidin et al. 2011), using infrastructure cost and value management expertise (Olanipekun, Aje & Abiola-Falemu 2013). In QSFs, employee Quantity Surveyors (QSs) provide the expertise, knowledge and skill relied upon for service delivery. This indicates their importance to the performance of QSFs, which is in line with Espejo (2000); Lawrence and Lorsch (1967).

One of the contexts in which organisations can be studied is their cultural effects. As opined by Zhang and Liu (2006), there are factors that seem to permeate organisational life and influence every aspect of organisation operation, and one of such is organisational culture (OC). According to Pandey (2014), OC lays the foundation of an organisation, builds and nurtures it, defines its purpose, sets its direction, prioritizes its tasks, guides its strategies and behaviour of its people and ultimately delivers its results. This implies OC forms an integral part of the general functioning of an organisation (Martins & Terblanche 2003). That is; OC epitomises the expressive character of an organisation. By definition, OC is a basic assumption, as well as the findings, invention, and development of an organisation dealing with external adaptation and internal integration (Wu & Lin 2013). This means that as groups evolve over time, they face two basic challenges: integrating individuals into an effective whole, and adapting effectively to the external environment in order to survive. As groups find solution to these problems over time, they engage in a kind of collective learning that creates the set of shared assumption and beliefs called culture (Ojo 2010).

The functionalist paradigm of OC views culture as a variable subject to conscious manipulation (Gajendran et al. 2012). In other words, OC is amenable to control. From this perspective it is possible to manage culture and to link culture to organisational performance, thereby implying a causal relationship (Gajendran et al. 2012). In corroboration, Pandey (2014), states that the study of OC gives solutions to most of organisational problems. In other words, a better understanding of the concept would allow people in organisations to solve problems and improve performance (Ojo 2010). Past research by Ramachandran (2013), Aldhuwaihi, Shee and Stanton (2012), Mathew, Ogbonna and Harris (2012), Su, Yang and Yang (2012), Dharmayanti, Coffey and Trigunarsyah (2011), Shaikh (2011), Ankrah (2007), Issa and Haddad (2007), Zhang and Liu (2006) and Lok and Crawford (2004) has explored and affirmed the potential of OC to solve problems within organisations.

Managerial and identity related challenges confront QSFs, and threatens their existence, growth and success (performance). In Nigeria, a developing country, the manifestations of these challenges are limited diffusion of the services of QSFs, inability of QSFs to directly secure jobs from clients, and inability of the QSFs to attract and retain quality QS personnel amongst others. It is important to address these problems so that QSFs can deliver high service performance in the Nigerian construction industry. Given the evidence exemplifying the possibility of addressing organisational challenges using OC, it is proposed that there is a relationship between the performance challenges of QSFs in Nigeria and their OC. It must be emphasised that OC exists in every organisation, including QSFs, whether noticed or not, desired or not, articulated or not (Pandey 2014; Ng & Kee 2013; Line 1999). Thus as a first step, this paper quantitatively elicits the OC in QSFs in Nigeria, prior to establishing relationship(s) with their performance in future research.

There are typology and dimension approaches to studying OC. Both are used to capture or represent important aspects of culture (Kessel, Oerlemans & Stroe-Biezen 2014). Though typologies are easier to comprehend, real cases correspond more with dimensions than typologies (Ankrah & Langford 2005; Hofstede 1997). Thus a dimension approach was adopted in this study, specifically the Organisational Culture Profile (OCP)'s business, people and environmental dimensions of OC. In terms of corresponding with real cases, the dimensions of OCP (Business, People and Environmental) relate well with the profit, client/employee and competitiveness orientation of QSFs respectively. Therefore in addition to eliciting the OC, the theoretical implications of the three dimensions for QSFs' performance challenges are given. That is, how does the OC in QSFs theoretically apply to solving the identified challenges? Findings of this research have implications for the management and employees of QSFs and Quantity Surveying researchers. The paper is structured as follows. First, a review of literature relevant to OC and the nature and challenges of QSFs is presented. Then the research method is described followed by the presentation of research findings.

Literature Review

Organisational Culture (OC)

OC has been described in different perspectives in the literature (Pandey 2014; Abiola-Falemu 2013; Shaikh 2011; Zhang & Liu 2006; Martins & Terblanche 2003). In Ng and Kee (2013) OC is defined as the collective programming of the mind which distinguishes the members of one organisation from another. In another definition, OC is the pattern of shared values and beliefs that help individuals understand organizational functioning and thus provide them with the norms for behaviour in the organization (Ramachandran 2013; Coffey, Willar & Trigunarsyah 2011). Taken together, OC is a distinguishing attribute (Pandey 2014; Rameezdeen & Gunarathna 2012; Aluko 2003), which is learned by people over the time period which they spend within the organization (Aftab, Rana & Sawar 2012; Liu & Fellows 2008), and eventually determines their perceptions and feelings and to some degree, their overt behaviour (Gajendran & Brewer 2012).

There are three levels or manifestation of OC: artefacts, values, and assumptions (Guevara 2014; Ng & Kee, 2013; Abiola-Falemu, Ogunsemi & Oyediran 2010; Parker & Bradley 2000). Artefacts are the visible structures and processes of an organisation, the architecture of its physical environment, its style, emotional displays, observable rites and ceremonies and, not least, its products (Guevara 2014; Igo & Skitmore 2005). Thus the manifestation of the artefact level OC in QSFs may be the Softwares for preparing Bills of Quantities (BOQ), the unique way of preparing and presenting BOQs or their project focus, which is largely on construction cost. However, artefact level does not provide deeper understanding of OC in QSFs (Guevara 2014). The second level of organisational culture comprises espoused beliefs, mental processes, knowledge and values (Barthorpe, Duncan & Miller 2000), where the strategy, goals and philosophy of an organisation are located, publicly justifying its existence, mission and measures (Guevara 2014). Assumptions, which are the unconscious taken-for-granted beliefs, perceptions, thoughts and feelings, are located at the third level of OC (Guevara 2014; Parker & Bradley 2000). Guevara (2014) however reiterates that contradictions showing at levels one and two can often be explained through the set of an organisation's implicit assumptions. In other words, assumption level of OC forms the ultimate source of organisational values and actions.

The philosophical conceptualisations of culture are mainly divided between the functionalist (culture as a variable) and non-functionalist (culture as a metaphor) paradigms (Gajendran & Brewer 2012; Gajendran et al. 2012; Maull, Brown & Cliffe 2001). The non-functionalist paradigm treats OC as a root metaphor whereby culture is more comprehensively described to explain human behaviour and its context, such that the behaviour becomes meaningful to an outsider (Gajendran et al. 2012). In this paradigm, reality is defined as subjective and multi-dimensional, with the possibility of different meanings attached to the same phenomenon (Wilson 2000). Also the philosophical stance of the non-functionalist paradigm of OC is more attuned to the qualitative research methodology (Gajendran et al. 2012). Conversely, the functionalist paradigm treats OC as something which may be influenced, changed and manipulated and in turn influences, changes and manipulates members and features of the organisation (Wilson 2000). In other words, basic values, assumptions and beliefs become enacted in established form of behaviours and activity that are reflected as structures, policies, practices, management practices and procedures (Gajendran et al. 2012; Martins & Terblanche 2003). This means outcomes e.g. performance, are embedded in the organisation's cultural symbols (values, assumptions, beliefs etc.) and when implemented, become the behaviour of members (See Guevara 2014; Aftab, Rana & Sarwar 2012; Ojo 2010; Martins & Terblanche 2003).

Thus OC is revealed in the actions and behaviour of employees (Abdul Halim et al. 2014) or conversely, in order to understand an organisation's observable behaviour patterns, it is crucial to understand its culture (Guevara 2014). Based on the behaviour (a reflection of OC) of employees in an organisation, customers or clients understand the culture of such organisation, and then form their own perceptions (Pandey 2014). Hence, if the customers' or clients' perception is positive, it is good for the organisations' market share, brand image etc., and if not, the organisation loses patronage (Pandey 2014). Another way in which OC influences employee behaviour towards performance is that culture improves employees' ties with one another (social embeddedness) (Kessel et al. 2014), and this increases the tendency of organisations to retain top performers and also attract new staff (Ojo 2010). Drawing on all these, OC is as a predictive and explanatory construct (Liu, Shuibo and Meiyung, 2006), that subscribes to the quantitative methodology (Gajendran et al. 2012; Maull, Brown & Cliffe 2001; Wilson 2000) or the functionalist paradigm.

In this paper, the notion is that through the functionalist paradigm, the three levels of OC (artefacts, values and assumptions) which embed organisational outcomes, when implemented, becomes the behaviour of employee QSs, and could be modified to improve the performance challenges in QSFs. Performance can be at the individual, work or process and organisational levels. However, individual and work performance levels translate to the organisational level (Aftab, Rana & Sarwar 2012), and thus the former is adopted in describing the performance of QSFs in this paper. Under certain circumstances the performance of an organisation is the account of how it progresses from state A to state B, the latter being in some way better than the former (Jirasinghe 2006).

Nature and Challenges of Quantity Surveying Firms (QSFs)

Quantity surveying firms (QSFs) are service oriented organisations providing cost and value management expertise on infrastructure procurement (Olanipekun, Aje & Abiola-Falemu, 2013; Abidin et al. 2011; Smith 2011; Fong & Choi 2006) in the construction industry. Predominantly QSFs rely on the skills, expertise, and knowledge of Quantity Surveyors (QSs) to satisfy clients' needs (Oyediran 2011; Nor, Mohamed & Egbu 2011; Fong & Choi 2006; Sonia 2005). This suggests the importance of people, or employee QSs to QSFs, and, as service based firms, their products are in the form of expert advice, services and consultancy (Abidin et al. 2011). In the past QSFs offered only traditional services such as valuation of the works in progress and settlement of final accounts (Pheng & Ming 1997). However, owing to competitive drives (Chong, Lee & Lim 2012; Smith 2004; Pheng & Ming 1997) and the need for survival and profitability (Ofori & Toor 2012; Abidin et al. 2010; Smith 2004), QSFs have expanded their scope of services (Zhou et al. 2012). The expanded services include among others, taxation advice, insurance valuations etc. (Smith 2011; Smith 2004).

Similar to other professional and management organisations, information technology (IT) is important to the success of QSFs (Smith 2004). For instance, BIM, if implemented in QSFs, will enable QSs to do work, or render their service more accurately and efficiently (Gee 2010). Though IT could be regarded as positive for QSFs, it also comes with some negatives (Smith 2011; Gee 2010; Smith 2004). Negatively, the use of BIM reduces the cost management task of QSs, almost making them irrelevant (Gee 2010). For QSFs the negativity and positivity that comes with IT is like being at cross roads on the decision to move from the past to the present (Ofori & Toor 2012). Thus the adoption of a technology such as BIM is low among UK's QSFs (Zhou et al. 2012), while less than half of the QSFs in Malaysia use measurement software application for Bill of Quantities (BOQ) preparation (Keng & Ching 2012). This appears like a problem for QSFs cutting across both developed and developing Nations.

QSFs are not without further challenges, most of which threatens their existence, growth and success (Frei & Mbachu 2013). According to Matipa, Kelliher and Keane (2009), consultancies tend to have a relatively small number of personnel, and Aliyu (2011), Smith (2011), Hardie et al. (2005) and Smith (2004) allude this assertion to QSFs. It is a challenge because QSFs will find themselves in dissonance to the logic of 'the bigger, the better' in business parlance. For instance, due to the smallness nature of QSFs, the research of Hardie et al. (2005) find that they do not innovate because of lack of money and time. Lack of time could be attributed to lack of adequate personnel to complete tasks on time. Other problems facing QSFs globally include the inability to deliver value for clients due to lack of value management knowledge (Bowen et al. 2010), and low diffusion of services rendered among clients due to poor marketing (Pheng & Ming 1997). One instance is Nigeria where many have still not come to terms with what quantity surveying as a profession is all about (Kadiri & Ayodele 2013), which is a question of identity (Onwusonye 2013).

In Australia and New Zealand, where Quantity Surveying practice is vibrant and active, QSFs' challenges include: fee cutting and bidding amongst firms, increased legal action due to professional indemnity insurance, the development of CAD, high level of QSFs' conservatism especially in terms of IT utilisation, incursion and encroachment of other professions, poor marketing and quality of graduates (Frei & Mbachu 2013; Smith 2004; Smith 2011). In Nigeria, a major problem of QSFs is that they tie their fortunes to friendly architecture and engineering firms because they cannot directly secure job commission from clients (Oyediran 2011). This practice has succeeded in shielding the identities of QSFs from private clients in Nigeria (Kadiri & Ayodele 2013). Other challenges faced by QSFs in Nigeria include weak business structure which no longer satisfies the present day business environment (Annunike 2011), low quality personnel due to the inability of the firms to train, motivate and retain specialist employees (Aliyu 2011), inadequacy of organisational resources e.g. stationary, resulting in low productivity (Ogunsemi, Oke & Awodele 2013), and outdated operational techniques (Atinuke 2010). In Nigeria, the problems of QSFs could be summarised into identity and organisational related, both of which may have contributed to the subsisting unimpressive delivery of Quantity Surveying practice in Nigeria (Onwusonye 2013; Fagbemi 2008).

In the literature there are few construction related researches focusing on OC. Majority of these researches evaluate the influence on OC on different social phenomenon in construction companies in different locations. Coffey's (2003) is company effectiveness (Hong Kong), Issa & Haddad's (2008) is knowledge sharing (USA), while Giritli et al.'s (2013) is leadership (Turkey). Oluwatayo, Amole and Adeboye (2014) and Albayrak and Albayrak (2014) only assess OC in descriptive terms in construction and architecture organisations respectively. The research of Liu and Fellows (2008) is the only research on OC in QSFs. The research investigates the effects of QSs' collective orientation (manifestation of culture) on their citizenship behaviour in various organisations. The findings of the research have implications only for QSs, which sharply contrasts the focus on QSFs in this paper. Therefore this paper contends that OC exists in QSFs, with theoretical implications for organisational challenges.

Conceptual Framework

In the literature review it was seen that culture is an effective variable that can be manipulated to enhance organisational outcomes. Also, there are challenges confronting QSFs, which have impacted on their performance and relevance. Thus this research asserts that OC is an effective tool to improve the organisational outcomes in QSFs. At first, and adopting the Dimensions of Organisational Culture (DOC), this research seeks to elicit the culture in QSFs and then describe in theoretical terms how it applies to the identified challenges. Though there are various theories, models or tools for understanding and diagnosing OC (Suppiah & Sandhu 2012), the Organisational Culture Profile (OCP) by Sarros et al. (2005) was adopted. The OCP has three dimensions; business, people and the environment. The three dimensions cover the areas that are important to service organisations like QSFs as they are business oriented (profit), environment oriented (competitiveness) and people oriented (employees and clients). These dimensions are described below.

Business Dimension of Organisational Culture

This dimension of OC emphasises goal accomplishment (Olanipekun 2012), achievement or competitiveness (Davies, Nutley & Mannion 2000), and innovativeness (Sarros et al. 2005). In other words, organisations with this dimension of OC expect their human resources to be competitive (Katamba 2010; Delobbe, Haccoun & Vandenberghe 2005), and quick to adapt quickly to opportunities (Alas, Ubius & Vanhala 2011). Further, such organisations allow openness to criticism, sharing of knowledge, sharing of information freely, encouragement of new ideas and taking risks (Khan et al. 2010). Consequently the organisations are productive, competitive and profitable (Alas et al. 2011).

People Dimension of Organisational Culture

The people dimension of OC has supportiveness and emphasis on reward as its sub cultures (Sarros et al. 2005). Supportiveness sub culture describes values or norms for interpersonal relationships and further, indicates the degree to which work activities are organized around teams rather than individuals (Katamba 2010). Reward sub culture emphasises the degree to which reward allocations are based on employee performance in contrast to seniority, favoritism or any other non-performance criterion (Katamba 2010). Thus the emphasis of people dimension of OC is concern for humanistic aspect in organisations (Vedina & Vadi 2007; Gray & Densten 2005; Davies et al., 2000). In manifesting the supportiveness sub culture, organisations do help provide training and counselling opportunities for their employees (Khan et al. 2010) which could lead to improved performance in line with Ankrah (2007). Also performance based reward promotes equity and competitiveness in organisations.

External Environment Dimension of Organisational Culture

Largely, this dimension of OC concerns for aspects in the external environment beyond the organization (Sarros et al. 2005). The sub cultures of this dimension are stability and social responsibility (Gray, Densten & Sarros 2003). In organisations, stability sub culture emphasises stability, execution of regulations and internal maintenance, and strives for consistency and control through clear tasks (Bashayreh 2009; Nel 2009). Also, social responsibility sub culture entails developing relationships with the society through positive contributions in the process of conducting organisational business. By implication, this dimension of OC determines the way in which organisational structure and support mechanisms contribute to the effectiveness of the organisation (means to achieve objectives), and focuses the image of the organisation to the outside world (whether it is a sought-after employer) (Mansor & Tayib 2010). Therefore it is more suited to larger organisations (Cameron 2004). In summary, when an organisation is balanced internally and externally, it is well positioned for success.

Research Methodology

Research Design, Instrument and Method of Analysis

The literature review above describes OC as a tool which can be used in organisations to improve outcomes. On this basis, this research investigates the dimensions of OC in QSFs in Nigeria. A survey method was used in eliciting necessary data for the study. Yin (2009) states that survey designs provide the best research method when prevalence and the incidence of a phenomenon are of interest. In this research, the interest is to investigate the pervasiveness of the dimensions of OC in QSFs with a view to giving their implications for addressing the identified challenges. Secondary data obtained through a literature review of relevant publications and information sourced from libraries and internet was used. The secondary data led to the tentative statement, and expressed using the OCP. The OCP provides the basis for measuring empirical research and was used in compilation of the questionnaire for the survey.

Thus in questionnaire compilation, the OCP is a standard questionnaire organised into three dimensions with the three dimensions further subdivided into seven sub-dimensions (e.g. *Business dimension*: innovation, performance orientation and competitiveness; *People dimension*: reward and supportiveness; *External environment dimension*: social responsibility and stability) and the seven sub-dimensions operationalised into four measurable items each, making a total of twenty eight. Since the OCP is a standard questionnaire, there was the need to modify the questions/measurables to engender the understanding of the respondents. For instance, a question/measurable under innovation is 'being innovative'. This was modified as: 'in this firm, innovative and creative ways of doing things are encouraged'. The manner of questioning in the questionnaire followed the rating of the extent of agreement with the cast measurables using 5-point Likert scale, with range of 5 as 'strongly agree' to 1 as 'strongly disagree'. In addition to the part of the questionnaire that contains the dimensions of OC, the general part elicited information on the hierarchy of respondents. Before sending out the questionnaire, it was tested for reliability using the Cronbach's Alpha test.   Cronbach's Alpha test measures internal consistency, which describes the extent to which all the items in a test measure the same concept or construct and hence it is connected to the inter-relatedness of the items within the test (Tavakol & Dennick 2011; Cortina 1993).

The population for this study is the database of QSFs in Lagos, Nigeria. Within the QSFs, the focus was on the three identified hierarchies of employees (Principal Partner, Senior, and Junior Quantity Surveyors). The identification and subsequent administration of questionnaires to these different hierarchies was to allow for robust and all inclusive response since perception of OC can vary among hierarchies in organisations. In support, Corley (2004) states that; one of the intriguing boundaries in which an organisation can be differentiated is organisational hierarchy and is seen to play significant role in determining the perceptions of organisational members. The total of 42 QSFs in Lagos Nigeria (data obtained from the Secretariat of the Nigerian Institute of Quantity Surveyors, Lagos State of Nigeria chapter) was sampled since they are within manageable size. In order to ensure that the questionnaires were administered to the identified hierarchies within the QSFs, purposive sampling was used i.e. the questionnaires were purposively given only to the QSs that are within each of the hierarchy. This implies at least three questionnaires were given per QSF.

Data Analysis

Prior to carrying out the data analysis, a reliability test was carried out to find out if it was reasonable to go ahead. Cronbach alpha test was carried out on the data and the result was as follows: Business dimension (performance orientation culture = 0.80; competitiveness culture = 0.81; innovation = 0.89), people dimension (reward culture = 0.82; supportiveness culture = 0.83) and environment dimension (social responsibility culture = 0.87; stability culture = 0.83). Given the above Cronbach alpha values, the research instrument was thus adjudged reliable based on the George and Mallery (2000) rule of thumb, where Cronbach's alpha (α) value ≥ 0.8 for research instruments was considered good. The analysis of the response on the agreement or disagreement on the dimensions of OC in QSFs was done using mean score. The mean score represents the average of the agreement among respondents on the questions being asked. To determine the direction of agreement of respondents, the mean score premise of decision of Johns (2010) was used. Johns (2010) stated that for a Likert scale that uses a 'decided' midpoint, the midpoint is a useful means of determining what might otherwise be a more or less random choice between agreement and disagreement. In this case, the mid-point of the 5-point Likert scale is 2.5, meaning that any measurable in this paper having a mean score ≥ 2.5 will be considered the agreed choice with respect to the question. Since there are three hierarchies of Quantity Surveyors, Kruskal-Wallis H test was carried to ascertain if there's significant difference in their opinion of the subject matter. The Kruskal-Wallis one-way analysis-of-variance-by-ranks test (or H test) is used to determine whether three or more independent groups are the same or different on some variable of interest when an ordinal level of data or an interval or ratio level of data is available (Chan & Walmsley 1997).

Firms' and Respondents' Details

Forty two QSFs in Lagos were sampled, with one questionnaire for each of the three identified hierarchies. This makes a total of one hundred and twenty six questionnaires administered, out of which ninety (71.42%) were returned and used for analysis. The breakdown of responses show that there were 21 Principal Partners (23.34%), 29 Senior Quantity Surveyors (32.22%) and 40 Junior Quantity Surveyors (44.44%). Notably, the response is skewed towards the lower hierarchy. This may not be unconnected with the fact that the responsibilities of higher hierarchies in most organisations, including QSFs are mostly outside the boundary of organisations. Such responsibilities include pursuing business opportunities and attending industry meetings. With this happening, there is less time devoted to issues they consider less important such as completing questionnaires.

However the 21 Principal Partner and 29 Senior Quantity Surveyor responses received could be counted upon for dependability, owing to the organisational knowledge they derive and exhibit in their positions. Across the three hierarchies of Quantity Surveyor respondents, 26.7%, 45.6% and 27.8% are academic holders of Higher National Diploma (HND), Bachelor and Master Degrees respectively. Professionally, 8 out of the 21 Principal Partners are Fellows of the Nigerian Institute of Quantity Surveyors (FNIQS). This is the highest professional qualification in Quantity Surveying in Nigeria, and this qualification is not awarded untill after more than 15 years of professional practice and service with unblemished record. This is an advantage to this research considering the professional exposure category. The other respondents (91.11%) are corporate members of the NIQS. Thus it could be inferred that the respondents who participated in the survey are both academically and professionally grounded, which should give credibility to the data collected.

The unit of analysis in this paper is the organisation, and thus it is important to focus on long standing organisations. It is logical to expect long standing organisations to have more robust OC than newly established. Thus out of the 42 QSFs, 18 were established over 15years ago, 15 had between 10-15 years in existence while the remaining 9 had about 5 years in existence. Given that the majority of the QSFs had longer years in existence, the respondents were asked if their respective QSFs had unified values, norms and practices that could be counted as their OC. All the 90 respondents from the 42 QSFs responded 'yes' The agreement of respondents on the existence of OC in QSFs may not be unconnected with the fact that the majority of firms have been in existence for a long time. Furthermore, all the QSFs have non-corporatised organisational structure (OS). 40 (95.24%) of the QSFs are partnerships, while the other two (4.76%) are sole-proprietorships. Since the focus of this research was not on organisation structure, and neither was the research dependent on the structure of QSFs, their non-corporatization would not impair the findings thereof.

Dimensions of Organisation Culture in Quantity Surveying Firms

The survey results of the dimensions of OC in QSFs are shown on Table 1. The business, people and external environment are the dimensions of OC in QSFs. Their group mean values are 4.35, 4.31 and 4.38 respectively. Obviously these mean score values are greater than the mid-point 2.50 threshold of John's (2010), implying that the respondents confirm by agreeing to have business, people and external environment as dimensions of OC in QSFs. That is, there are three dimensions of OC in QSFs. The analysis was done in line with the operationalisation of each of the three dimensions of organisational culture into distinct seven sub-cultures. For the business dimension, the mean scores for competitiveness sub culture by Principal Partner (PP), Senior Quantity Surveyor (SQS) and Junior Quantity Surveyor (JQS) are 4.48, 4.79 and 4.45 respectively. For the performance orientation sub culture, the mean scores by PP, SQS, and JQS are 4.19, 4.28 and 4.25 while their mean values for innovation sub culture are 4.38, 4.10 and 4.13 respectively. The people dimension has reward and supportiveness as its sub-cultures, where the mean values for both by the PP, SQS and JQS are 4.43, 4.03, 4.05 and 4.43, 4.34, 4.00 respectively. For the external environment dimension, the mean values for stability sub culture by PP, SQS and JQS are 4.48, 4.48 and 4.25 respectively while their mean values for social responsibility sub culture are 4.43, 4.45 and 4.20 respectively. From Table 1, it could be seen that the mean values for the dimensions of OC and their sub-dimensions are ≥4.00. The high mean values suggest that culture not only exist, but are pervasive, in QSFs.

The Principal Partner (PP), Senior Quantity Surveyor (SQS) and Junior Quantity Surveyor (JQS) are the hierarchies in QSFs, and except for the supportiveness sub culture, there is insignificant difference in their opinions regarding the dimensions of OC at 5% significance level (Table 1). This implies that these hierarchies are unified in their views, which is an advantage to pursuing organisational goals.

Table 1: Dimensions of organisational culture

Dimensions of OC		PP		SQS		JQS		BD	PD	ED	GMR	Chi-square	Sig.
		MS	Rank	MS	Rank	MS	Rank						
Business Dimension	Performance orientation culture	4.19	3	4.28	2	4.25	2					0.760	0.684
	Competitiveness culture	4.48	1	4.79	1	4.45	1					3.005	0.223
	Innovation culture	4.38	2	4.10	3	4.13	3					2.931	0.231
	GROUP MEAN SCORE							4.35			2nd		
People Dimension	Reward culture	4.43	1	4.03	2	4.05	1					3.912	0.141
	Supportiveness culture	4.43	1	4.34	1	4.00	2					12.553	0.002*
	GROUP MEAN SCORE								4.31		3rd		
Environment Dimension	Stability culture	4.48	1	4.48	1	4.25	1					2.849	0.241
	Social responsibility culture	4.43	2	4.45	2	4.20	2					2.049	0.231
	GROUP MEAN SCORE									4.38	1st		*Significant @5% level

OC - Organisational Culture
PP - Principal Partner, SQS - Senior Quantity Surveyor, JQS - Junior Quantity Surveyor
MS - Mean Score, GMR - Group Mean Rank
BD - Business Dimension, PD - People Dimension, ED - Environment Dimension

Conclusion

This paper investigated the DOC in QSFs in Nigeria. It finds that the business, people and the external environment are the DOC in QSFs. This supports the fact that every organisation has a culture (Pandey 2014; Ng & Kee 2013; Line 1999). However, the problems or challenges confronting QSFs suggest that the importance of their OC has not been appreciated, and therefore not deployed to address the problems. Taking into account the identity and management related problems of QSFs in Nigeria, it may be concluded that OC is a relevant solution.

The nature of identity problem of QSFs is that their services are not well diffused among relevant stakeholders e.g. clients, policy makers and majority of the public, and therefore not recognised as the professional organisation responsible for cost management in Nigeria. As stated by Hatch and Schultz (1997), the identity of an organisation is expressed through its OC, which is subject to interpretation by others. Also an organization's identity is comprised of its member perceptions, its material features and its actions (Tyworth 2014). By implication, the identity or the diffusion of the services offered by QSFs in Nigeria is a function of the OC exhibited in them.

Specifically, this will affect both the business and people dimensions of OC in QSFs. Notably, the business dimension of OC emphasises on results and competitiveness oriented performance from employees (e.g. Katamba 2010; Delobbe, Haccoun & Vanderberghe 2005; Davies et al. 2000), while the people dimension of OC is about humanistic concerns such that the personnel in an organisation are motivated to deliver performance (e.g. Khan, Usoro & Majewski 2010; Vedina & Vadi 2007). Thus both dimensions of OC operate interdependently. Also, relating the interdependence of both dimensions of OC to the identity problem of QSFs has implications for employee QSs. That is, how QSs in QSFs portray their organisation to the public, and how they contribute to the service quality of the organisation. Both positive perception and quality contribution to service quality are a function of the welfare and support provided by the management to QSs. It is logical for QSs, who are well remunerated and gets their voice heard in organisational matters, to constantly heap praises for the organisation (people dimension of OC).

It is also imaginable to expect QSs who are supported with an exquisite and result oriented work environment to deliver high performance (business dimension of OC). These lead to what is called competitive workforce, which are personnel that are well motivated to deliver acme performance. This is an attraction to clients or customers, and the society at large. At this point QSFs are no longer obscure or lacking identity. In construction generally, such organisations are quick to get job commissions from clients because of their capacity to retain quality and motivated workforce. Invariably the organisational performance is improved when there are job commissions and the workforce to deliver them.

Weak organisational structure, outdated operational and inefficient service delivery techniques are the nature of organisational problems identified in QSFs in Nigeria. As in other locations, QSFs in Nigeria are also bedevilled by technological and financial shortcomings. All these can be summed up as management related problems that affect organisational performance. The functionalist paradigm of OC purports that culture can be controlled by managers in organisation. Within this paradigm lies the notion that OC is instrumental to organisational performance (Fellows 2006). Therefore by implication, the dimensions of OC are important to solving the organisational or management related problems in QSFs. It is thus incumbent upon managers of QSFs to operate through the lens of their OC. In fact, strategic management theory demands that the culture of organisations be held as an important element in the formulation of any strategy (Barthorpe et al. 2000). Based on this, it could be said that the prevention of management or organisational related problems in QSFs lies in the consideration of their OC by managers, when management actions are taken.

For instance, an organisation structure (OS) capable of driving the performance of organisations is one that is dynamic and flexible (Sadeghian, Kafashpoor & Lagzian 2013). That is, OS that can adapt better to the needs of the environment. Among the dimensions of OC, it could be said that implementing OS in QSFs should be in cognizance with the external dimension of OC. Inferring from Nel (2009), taking such cognizance prevents discord with the environment in which the QSF is located. Therefore, instead of weakened OS (Annunike 2011), managers of QSFs can implement a performance enhancing OS by giving proper consideration to OC. The same applies to other management related problems in QSFs, especially the financial related. As an instrument for improving organisational performance, it is well documented in the literature how OC can be adopted to improve financial performance (Yusoff 2011; Davidson, Melinde & Delene 2007). Thus it could be said that the attaining of competitive financial posture in the QSFs is a function of their OC.

The findings of this research have implications for the management and employees of QSFs and Quantity Surveying researchers. The implications affect employee QSs and managers of QSFs, and the Quantity Surveying academia who have evidently neglected the organisational aspect of the Quantity Surveying profession. The outcome of this study, suggests that for best performance, managers of QSFs in Nigeria should give cognizance to OC when taking management decisions or actions. This approach will not only help in mitigating the existing problems but prevent future organisational performance failures.

Second, the Quantity Surveying academia should direct research focus on various aspects of QSFs, as this has the tendency of providing seminal solutions to recurring problems. This paper adopts survey research design by using the OCP to investigate, and as a basis for describing the OC in QSFs. The possibility remains that the modified OCP questionnaire might not fully capture the OC in QSFs. A mixture of survey and case study (e.g. participant observation) would have enriched the data used in this paper. Thus this paper points to the need for further enquiries using a mixed a research design in studying the OC in QSFs. Also, investigating the relationship between the performance and OC in QSFs will be a significant research.

References

Abdul Halim, H.A., Ahmad, N.H., Ramayah, T. & Hanifah, H. 2014, 'The Growth of Innovative Performance among SMEs: Leveraging on Organisational Culture and Innovative Human Capital', Journal of Small Business and Entrepreneurship Development, 2 (1), 107-25.

Abidin, N.Z., Yusof, N., Hassan, H. & Adros, N. 2010, 'Applying Competitive Strategy in Quantity Surveying firms: An Evolving Process', Asian Journal of Management Research, 2 (1), 61-73.

Abiola-Falemu, J.O., Ogunsemi, D.R. & Oyediran, O.S. 2010, 'Assessment of Organisational Culture and Innovation Practices of Construction Companies in Southwest Nigeria', In CIB Task Group and Working Commission, P. Barret et al., (ed) TG59 & W112-Special Track 18th CIB World Building Congress, May 2010, Salford, United Kingdom, 218-33.

Abiola-Falemu, J.O. 2013, 'Organisational Culture, Job Satisfaction and Commitment of Lagos-based Construction Workers', Journal of Business and Management, 13 (6), 108-20.

Aftab, H., Rana, T. & Sarwar, A. 2012, 'An Investigation of the Relationship between Organizational Culture and the Employee's Role Based Performance: Evidence from the Banking Sector', International Journal of Business & Commerce, 2 (4), 1-13.

Alas, R., Ubius, U. & Vanhala, S. 2011, 'Connections between Organisational Culture, Leadership and the Innovation Climate in Estonian Enterprises', In E-Learning Bachelor's, International Prestige. E-Leader, January 2011, Vietnam, 1-15.

Albayrak, G. & Albayrak, U. 2014, 'Organizational Culture Approach and Effects on Turkish Construction Sector', In 5th International Conference on Chemical, Biological and Environmental Engineering - ICBEE & 2nd International Conference on Civil Engineering – ICCEN, D. Yang, (ed) APCBEE Procedia, September 2013, New Delhi, India, 252-57.

Aldhuwaihi, A., Shee, H.K. & Stanton, P. 2012, 'Organisational Culture and the Job Satisfaction-Turnover Intention Link: A Case Study of the Saudi Arabian Banking Sector,' World, 2 (3), 127-41.

Aliyu, M. 2011, 'Need for Specialisations / Faculties in Quantity Surveying Practice,' In Quantity Surveying and the Anti- Corruption Crusade-Achieving Value for Money in Project Cost in Nigeria, Quantity Surveying Assembly and Colloquium, Septmeber 201,1 Abuja, Nigeria, 11-28.

Aluko, M.A.O. 2003, 'The Impact of Culture on Organizational Performance in Selected Textile Firms in Nigeria', Nordic Journal of African Studies, 12 (2), 164-79.

Ankrah, N.A., & Langford, D.A. 2005, 'Architects and contractors: a comparative study of organizational cultures,' Construction Management and Economics, 23 (6), 595-607.

Ankrah, N.A. 2007, 'An Investigation into the Impact of Culture on Construction Project Performance,' Doctoral Thesis submitted to the University of Wolverhampton, UK.

Annunike, E.B. 2011, 'The changing roles of the Quantity Surveyor in National Development', In Quantity Surveying and the Anti- Corruption Crusade-Achieving Value for Money in Project Cost in Nigeria, Quantity Surveying Assembly and Colloquium, Septmeber 2011, Abuja, Nigeria, 1-10.

Atinuke, J.O. 2010, 'Construction cost data management by quantity surveying firms in Nigeria', In West Africa Built Environment Research (WABER) Conference, S. Laryea et al., (ed), WABER Publishing, July 2010, Accra, Ghana, 247-54.

Barthorpe, S., Duncan, R. & Miller, C. 2000, 'The pluralistic facets of culture and its impact on construction',   Property   Management,   18   (5),   335-51.

Bashayreh, A.M. 2009, 'Organizational culture and job satisfaction: A case of academic staffs at Universiti Utara Malaysia (UUM)', Masters Degree thesis, University Utara Malaysia (UUM). Available from: http://etd.uum.edu.my/1632/ 20 November, 2014.

Bowen, P., Cattell, K., Edwards, P. & Jay, I. 2010, 'Value management practice by South African quantity surveyors', Facilities, 28 (1/2), 46-63.

Cameron, J.E. 2004, 'A three-factor model of social identity,' Self and Identity, 3 (3), 239-62. doi:

Chan, Y. & Walmsley, R.P. 1997, 'Learning and understanding the Kruskal-Wallis one-way analysis-of-variance-by-ranks test for differences among three or more independent groups,' Physical Therapy, 77 (12), 1755-61.

Chong, B.L., Lee, W.P. & Lim, C.C. 2012, 'The Roles of Graduate Quantity Surveyors in the Malaysian Construction Industry,' International Proceedings of Economics Development & Research, 37, 17-20.

Coffey, V. 2003, 'The organisational culture and effectiveness of companies involved in public sector housing construction in Hong Kong', In Professionalism in Construction: Culture of High Performance, A. Liu & Fellows, R., (eds), CIB TG 23 International Conference, October 2003, Hong Kong, 27-43.

Coffey, V., Willar, D. & Trigunarsyah, B. 2011, 'Profiles of organisational culture in Indonesian construction companies', In the Sixth International Structural Engineering and Construction Conference, S.O. Cheung, (ed), Modern Methods and Advances in Structural Engineering and Construction, June 2011, Zurich, Switzerland. Available from: http://eprints.qut.edu.au/41063/ 20 November 2014.

Corley, K.G. 2004, 'Defined by our strategy or our culture? Hierarchical differences in perceptions of organizational   identity   and   change',   Human   Relations,   57   (9),   1145-77.

Cortina, J.M. 1993, 'What is coefficient alpha? An examination of theory and applications', Journal of Applied Psychology, 78 (1), 98.

Davidson, G., Melinde C. & Deléne V. 2007, 'Organisational culture and financial performance in a South African investment bank', SA Journal of Industrial Psychology, 33 (1), 38-48.

Davies, H.T.O., Nutley, S.M. & Mannion, R. 2000, 'Organisational Culture and Quality of Health Care', Quality in Health Care, 9, 111–19.

Delobbe, N., Haccoun, R.R. & Vandenberghe, C. 2005, 'Measuring core dimensions of organizational Culture: A Review of Research and Development of a New Instrument', retrieved 30 October 2012 http:// www.uc llouvain. be/ cps/ucl/doc /iag/ documents/ WP_53_ Delobbe.pdf

Dharmayanti, G., Coffey, V. & Trigunarsyah, B. 2012, 'The impact of organisational culture on project selection: what is the appropriate culture type?', In Advancing Civil, Architectural and Construction Engineering & Management, Third International Conference on Construction In Developing Countries (ICCIDC–III), July 2012, Bangkok, Thailand, 109-15.

Espejo, R. 2000, 'Self-construction of desirable social systems', Kybernetes, 29 (7/8), 949-63.

Fagbemi, A.O. 2008, 'Assessment of Quantity Surveyors' service quality in Lagos state, Nigeria', M.Tech Thesis Dissertation, Unpublished, Quantity Surveying Department, Federal University of Technology, Akure, Nigeria.

Fellows, R. 2006. Understanding approaches to culture. Construction Information Quarterly, 8 (4), 159-66.

Fong, P.S., & Choi, S.K.A. 2006, 'A framework of knowledge processes for professional quantity surveying firms in Hong Kong', In Joint International Conference on Computing and Decision Making in Civil and Building Engineering, June 2006, Montréal, Canada, 268-77.

Frei, M., Mbachu, J. & Phipps, R. 2013, 'Critical success factors, opportunities and threats of the cost management profession: the case of Australasian quantity surveying firms', International Journal of Project Organisation and Management, 5 (1), 4-24.

Gajendran, T. & Brewer, G. 2012, 'Cultural consciousness and the effective implementation of information and communication technology', Construction Innovation: Information, Process, Management, 12 (2), 179-97.

Gajendran, T., Brewer, G., Dainty, A.R. & Runeson, G. 2012, 'A conceptual approach to studying the organisational culture of construction projects', Australasian Journal of Construction Economics and Building, 12 (2), 26.

Gee, C. 2010, 'The influence of buidling information modelling on the quantity surveying profession', retrieved from http://repository.up.ac.za/handle/2263/16349 on 1 October, 2014.

George, D. & Mallery, P. 2000, SPSS for Windows Step by Step: A Simple Guide and Reference 9.0 Update, 2nd ed., Allyn and Bacon, Boston.

Giritli, H., Öney-Yazıcı, E., Topçu-Oraz, G. & Acar, E. 2013, 'The interplay between leadership and organizational culture in the Turkish construction sector', International Journal of Project Management, 31 (2), 228-38.

Gray, J.H., Densten, I.L., & Sarros, J.C. 2003, 'Size Matters: Organisational Culture in Small, Medium and Large Australian organisations', Journal of Small Business and Entrepreneurship, 17 (1), 42-53.

Gray, J.H. & Densten, I.L. 2005, 'Towards an Integrative Model of Organizational Culture and Knowledge Management', International Journal of Organisational Behaviour, 9 (2), 594-603.

Guevara, B. 2014, 'On methodology and myths: exploring the International Crisis Group's organisational culture', Third World Quarterly, 35 (4), 616-33.

Hardie, M.P., Miller, G., Manley, K. & McFallan, S. 2005, 'The quantity surveyor's role in innovation generation, adoption and diffusion in the Australian construction industry', In The Queensland

University of Technology Research Week International Conference, A. Sidwell, (ed), Conference Proceedings, Queensland University of Technology, July 2005 Brisbane, Australia.

Hatch, M.J., & Schultz, M. 1997, 'Relations between organizational culture, identity and image. European Journal of marketing', 31 (5/6), 356-65.

Hofstede, G. 1997, 'Cultures and Organizations: software of the mind', McGraw-Hill, New York.

Igo, T. & Skitmore, M. 2006, 'Diagnosing the organizational culture of an Australian engineering consultancy using the competing values framework', Construction Innovation: Information, Process, Management, 6 (2), 121-39.

Issa, R.R., & Haddad, J. 2008, 'Perceptions of the impacts of organizational culture and information technology on knowledge sharing in construction', Construction innovation: Information, process, management, 8 (3), 182-201.

Jirasinghe, E.H. 2006, 'Managing & Measuring Employee Performance' Kogan Page, London.

Johns, R. 2010, 'Survey question bank: Methods Fact Sheet 1, Likert items and scales', University of Strathclyde.

Kadiri, D.S. & Ayodele, E.M. 2013, 'Constraints to Quantity Surveying Awareness in Nigeria', Civil and Environmental Research, 3 (11), 17-21.

Katamba, D. 2010, 'Corporate Social Responsibility, Organizational Culture, Ethical citizenship and Reputation of Financial Institutions in Uganda', Master Degree Thesis submitted to the School of Graduate Studies, Makerere University, Uganda.

Keng, T.C. & Ching, Y.K. 2012, 'A study on the use of measurement software in the preparation of bills of quantities among Malaysian quantity surveying firms', In the Ninth International Conference on ICT and Knowledge Engineering, Institute of Electrical Engineers, Piscataway, N.J, January 2011, Bangkok, Thailand, 53-58.

Kessel, F.G., Oerlemans, L.A. & van Stroe-Biezen, S.A. 2014, 'No creative person is an island: organisational culture, academic project-based creativity, and the mediating role of intra-organisational social ties', South African Journal of Economic and Management Sciences, 17 (1), 46-69.

Khan, U.I., Usoro, A. & Majewski, G. 2010, 'An Organisational Culture Model for Comparative Studies: A Conceptual View', International Journal of Global Business, 3 (1), 53-82.

Lawrence, P.R., & Lorsch, J.W. 1967, 'Differentiation and integration in complex organizations', Administrative science quarterly, 12 (1), 1-47.

Line, M.B. 1999, 'Types of Organisational Culture', Library Management, 20 (2), 73-5.

Liu, A.N.M., Shuibo, Z. & Meiyung, L. 2006, 'A framework for assessing organisational culture of Chinese construction enterprises', Engineering, Construction and Architectural Management, 13 (4), 327-342.

Liu, A.M. & Fellows, R. 2008, 'Behaviour of quantity surveyors as organizational citizens' Construction Management and Economics, 26 (12), 1271-82.

Lok, P. & Crawford, J. 2004, 'The effect of organisational culture and leadership style on job satisfaction and organisational commitment: A cross-national comparison', Journal of Management Development, 23 (4), 321-38.

Mansor, M. & Tayib, M. 2010, 'An Empirical Examination of Organisational Culture, Job Stress and Job Satisfaction within the Indirect Tax Administration in Malaysia' International Journal of Business and Social Science, 1 (1), 81-95.

Martins, E.C, & Terblanche, F. 2003, 'Building organisational culture that stimulates creativity and innovation', European Journal of Innovation, 6 (1), 64-74.

Mathew, J., Ogbonna, E. & Harris, L.C. 2012, 'Culture, employee work outcomes and performance: An empirical analysis of Indian software firms', Journal of World Business, 47 (2), 194-203.

Matipa, W.M., Kelliher, D. & Keane, M. 2009, 'A strategic view of ICT supported cost management for green buildings in the quantity surveying practice', Journal of Financial Management of Property and Construction, 14 (1), 79-89.

Maull, R., Brown, P. & Cliffe, R. 2001, 'Organisational culture and quality improvement', International Journal of Operations & Production Management', 21 (3), 302-26.

Nel, L.J. 2009, 'Shared values and organisational culture, a source for competitive advantage: a comparison between Middle East, Africa and South Africa using Competitive Values Framework', published Master Degree dissertation submitted to the Global Institute of Business Science, University of Pretoria.

Ng, H.S. & Kee, D.M.H. 2013, 'Organisational Culture can be a Double-edged Sword for Firm Performance', Research Journal of Business Management, 7 (1), 41-52.

Nor, F., Mohamed, O. & Egbu, C. 2011, 'Knowledge sharing initiatives in quantity surveying firms in Malaysia: Promoting, inhibiting and challenge factors', In: Egbu, C. and Lou, E.C.W. (eds) Procs 27th Annual ARCOM Conference, 5-7 September 2011, Bristol, UK, Association of Researchers in Construction Management, 593-601.

Ofori, G. & Toor, S. 2012, 'Role of leadership in transforming the profession of quantity surveying', Australasian Journal of Construction Economics and Building, 9 (1), 37-44.

Ogunsemi, D.R., Awodele, O.A. & Oke, A.E. 2013, An Examination of the Management of Quantity Surveying Firms in Nigeria, delivered at the 2013 Annual Conference of Registered Quantity Surveyors with theme: 'Quantity Surveying Profession: Unbundling Latent Competencies and Developing New Frontiers' Thursday 26th and Friday 27th September, Abuja.

Ojo, O. 2010, 'Organisational Culture and Corporate Performance: Empirical Evidence from Nigeria', Journal of Business System, Governance and Ethics, 5 (2), 1-12.

Olanipekun, A.O. 2012, 'Effects of Organisational Culture on the Performance of Quantity Surveying Firms in Lagos, Nigeria' Master Degree Thesis submitted to the School of Postgraduate Studies, Federal University of Technology, Akure, Nigeria.

Olanipekun, A.O., Aje, I.O. & Abiola-Falemu, J.O. 2013, 'Effects of Organisational Culture on the Performance of Quantity Surveying firms in Nigeria', International Journal of Humanities and Social Science, 3 (5), 206-15.

Oluwatayo, A.A., Amole, D. & Adeboye, A.B. 2014, 'Architectural Firms in Nigeria: A Study of Organizational Culture and Determinants', Global Journal of Researches in Engineering: Civil And Structural Engineering, 14 (1), 12-26.

Onwusonye, S.I.J. 2013, 'Quantity Surveying Profession and the Identity Crises', In Quantity Surveying Profession: Unbundling Latent Competencies and Developing New Frontiers, Annual Conference of Registered Quantity Surveyors, September 2013 Abuja, Nigeria.

Oyediran, O.S. 2011, 'Challenges to Efficient Service Delivery by Quantity Surveyors', In Quantity Surveying and the Anti- Corruption Crusade-Achieving Value for Money in Project Cost in Nigeria, Quantity Surveying Assembly and Colloquium, September 2011, Abuja, Nigeria, 77-86.

Pandey, P. 2014, 'Organisational culture-a root to prosperity', Management Insight, 10 (1), 74-80.

Parker, R. & Bradley, L. 2000, 'Organisational culture in the public sector: evidence from six organisations', International Journal of Public Sector Management, 13 (2), 125-41.

Pheng, L.S. & Ming, K.H. 1997 'Formulating a strategic marketing mix for quantity surveyors', Marketing Intelligence & Planning, 15 (6), 273-80.

Ramachandran, S. 2013, 'A peep into organisational culture', Middle-East Journal of Scientific Research, 15 (12), 1925-27.

Rameezdeen, R. & Gunarathna, N. 2012, 'Organisational culture in construction: an employee perspective', Australasian Journal of Construction Economics and Building, 3 (1), 19-30.

Sadeghian, S., Kafashpoor, A. & Lagzian, M. 2013, 'Assessing Impact of Organizational Culture and Organizational Structure on Organizational Effectiveness through Knowledge Management Case Study: Mashhad's Science and Technology Park Author's Details', International Journal of Management Sciences & Business Research, 2 (12), 24-31.

Sarros, C.J., Gray, J., Densten, I.L. & Cooper, B. 2005, 'The Organizational Culture Profile Revisited and Revised: An Australian Perspective', Australian Journal of Management, 30 (1), 159-82.

Shaikh, S. 2011, 'The Study on the Relationship between Organisational Culture and Knowledge Management', Journal of Money, Investment and Banking, 19, 21-28.

Smith, P. 2004, 'Trends in the Australian Quantity Surveying Profession 1995-2003', Retrieved from http://epress.lib.uts.edu.au/research-publications/handle/10453/7332 on 30 September 2014.

Smith, P. 2011, 'Information Technology and the QS Practice', Australasian Journal of Construction Economics and Building, 1 (1), 1-21.

Sonia, K.Y.S. 2005, 'A framework of knowledge processes for professional quantity surveying firms in Hong Kong', Retrieved from http://repositor y.lib.polyu. edu.hk/jspui /handle/ 10397/3 400 on 1 October 2014.

Su, Z., Yang, D. & Yang, J. 2012, 'The match between efficiency/flexibility strategy and organisational culture', International Journal of Production Research, 50 (19), 5317-29.

Suppiah, V. & Sandhu, M.S. 2012, 'Organisational culture's influence on tacit knowledge-sharing behaviour', Journal of Knowledge Management, 15 (3), 462-77.

Tavakol, M. & Dennick, R. 2011, 'Making sense of Cronbach's alpha', International journal of medical education, 2, 53-55.

Tyworth, M. 2014, 'Organizational identity and information systems: how organizational ICT reflect who an organization is', European Journal of Information Systems, 23 (1), 69-83. doi:

Vedina, R. & Vadi, M. 2007, 'A National Identity Perspective on Collectivistic Attitudes and Perception of Organisational Culture', Baltic Journal of Management, 3 (2), 129-44.

Wilson, E. 2000, 'Inclusion, exclusion and ambiguity–The role of organisational culture', Personnel Review, 29 (3), 274-303.

Wu, Y.H. & Lin, M.M. 2013, 'The relationships among business strategies, organisational performance and organisational culture in the tourism industry', South African Journal of Economic and Management Sciences, 16 (5), 1-8.

Yin, R.K. 2009, 'Case Study Research: Design and Methods' Sage Publications, California.

Yusoff, W.F.W. 2011, 'Does Organizational Culture Infleunce Firm Performance in Malaysia?', International Journal of Multidisciplinary Research, 1 (3), 1-12.

Zhang, S.B., & Liu, A.M. 2006, 'Organisational culture profiles of construction enterprises in China', Construction Management and Economics, 24 (8), 817-28.

Zhou, L., Perera, S., Udeaja, C. & Paul, C. 2012, 'Readiness of BIM: a case study of a quantity surveying organisation', In: First UK Academic Conference on BIM, D. Greenwood, (ed.), Newcastle Business School & School of Law, September 2012 Newcastle upon-Tyne, UK, 118-28.

Management Challenges within Multiple Project Environments: Lessons for Developing Countries

9

Noor Ismah Hashim, Nicholas Chileshe, and Bassam Baroudi
(University of South Australia, Australia)

Abstract

In the construction industry, multiple project environments (MPE) exist where more than one project is managed simultaneously. The driving force behind MPEs is the pragmatic allocation of resources encumbered by uncertain economic times. However, MPEs create management challenges that need to be addressed. For that reason, this paper aims to investigate the challenges in respect to managing MPEs within the construction industry. It essentially reviews state-of-art knowledge in respect to MPEs identifying the rationale behind their development. At this stage it would appear that the interdependency and uncertainty within inputs, processes and outputs are major contributing factors to the MPE problem. It is of note that the majority of these findings were based within the context of developed countries. Hence, this review sets out to inform practitioners from developing countries in respect to lessons learned within more developed countries. This review is expected to lead to further investigations on MPEs and their inherent challenges.

Keywords: Construction industry, project management, multiple project environments

Introduction

Organisations are taking management initiatives by shifting the paradigm of project management to the management of multiple projects (Blomquist and Müller, 2006, Pennypacker and Dye, 2002, Evaristo and van Fenema, 1999) as an efficient vehicle to successfully deliver improvements and changes due to the unpredictable economic climate (Shehu and Akintoye, 2010). For the construction industry, it needs to assimilate new steps to intervene with such uncertainties to survive. Thus managers are altering their strategic direction to expand opportunities and expand capacity for marketing, sourcing, introducing new infrastructure and taking advantage of distributed location (Dooley et al., 2005).

Studies on the management of multiple projects are dominated by the high technology industry (Caniëls and Bakens, 2011, Patanakul and Milosevic, 2009, Maylor et al., 2006) specifically on the new product development. Few studies have examined and little analytical attention have been paid to the management of multiple projects environment within the construction industry (Gholipour, 2006, Blismas et al., 2004, Dubois and Gadde, 2002).

Furthermore, most studies have demonstrated the existence of multiple projects environment from the context of developed country. Although studies on construction industry in the context of developing country has been acknowledged in the literature (Ngowi, 2002, Ofori, 2000), little recognition has been given to the multiple project environments within the construction industry. It is important to recognise the management of multiple projects environment from the developing country because the construction industry among countries is different as presented in the cultural studies of the construction projects, firm and site by Baiden and Price (2011). Thus, the complexity of challenges will be different in the level of socio-economic stress, chronic resource shortages, institutional weaknesses and a general inability to deal with the key issues (Ofori, 2000).

Therefore, this review builds on and contributes to the work in the multiple projects environment (MPE) within the construction industry. Although studies in the MPE have examined the development of effectiveness in management (Chinowsky et al., 2011, i.e

Patanakul and Milosevic, 2009), there has been lack of an extended study on the challenges that hinder the effectiveness in managing the MPE. This paper will address this issue by reviewing how the assertions about challenges attributed to the MPE within the construction industry have been transformed into lessons to be learned for the developing countries. Consequently, this review provides additional insight into the constructive processes of exploring challenges by explicating the challenges through which the uncertainty and interdependency is constructed from the complexity of management.

Multiple project environments

Multiple project environments (MPEs) have been defined in many ways in the research. To describe the management of MPEs, studies have been premised with terms such as multi-project, portfolio, programme, macro-project, mega-project, giving the impression of similar meanings (Turner, 2009, Project Management Institute, 2008). The inconsistency in definition has led to limited insights due to confusion and diverse understanding (Shehu and Akintoye, 2010, Milosevic, 2009) into the relationship of MPEs and their challenges. The definition of the MPE in this review reveals some features that best describe the nature of the construction industry. Initially, MPEs was referred to, "an organisational level environment in which multiple projects are managed concurrently" (Patanakul and Milosevic, 2009, p. 217). However, this definition needs to be extended not to focus only on more than one project managed simultaneously, but also at various locations (Evaristo and van Fenema, 1999), on the possibility of involvement from multiple organisations (Dubois and Gadde, 2002)

These two features of multiple projects at various locations and involves multiple organisations are important in defining MPEs. The first feature stressed on various locations because within the construction industry, projects are influenced by geographical location which includes international and domestic distribution whether in a local region or elsewhere. This distribution is due to the potential benefits of the physical location and where professionals are involved in the project operation location (Zavadskas et al., 2004). One project can be performed in several sites concurrently, as long as the correspondent actions share the same objectives (Evaristo and van Fenema, 1999). The management of these projects is assumed to be either centralised or distributed located in any of the sites or nodes. The challenge of project's location of multiple projects is related to the focus on the co-ordination mechanisms, with the option of either focusing on inter-site or boundary spanning across sites, or concentrate on intra-site or boundary spanning across projects (Hashim and Chileshe, 2012).

The second feature originated from the construction management which is complicated by several organisations involved in the supply chain. The organisations are also engaged in other projects in which they have to coordinate their activities and resources with different sets of organisations. This affiliation shows that an organisation is capable in managing more than one project simultaneously in the construction industry (Dubois and Gadde, 2002) and supports project-based structures (Söderlund, 2004). The increased use of project-based structures defines the nature of multiple project environments with the involvement of multi-project organisations.

From these features, the representation of challenges instigated from the complexity in managing multiple projects could be illustrated. For example, the projects located in multiple locations will focus on the co-ordination mechanisms, on single unit without segregating the projects into multiple units in sharing the projects goal and objectives even though they are widely distributed from each other (Desouza and Evaristo, 2004). On the other hand, projects which involves with multiple organisations will easily create conflict between the team mates, and impede the establishment of "organisation culture" of multiple projects environment particularly between different levels of management or between other projects, especially when competing after the same resources (Fricke and Shenbar, 2000, Olford, 2002). Therefore, these features illustrate the challenges in managing the MPE that will minimise the effectiveness in managing the projects. .

Methods

The focus of this paper is given to the research on the multiple projects management within the developed country. Therefore, little consideration will be given to the contributions within traditional project management, such as project risk management, project planning and etc. The attention is on studies and articles that analyse the complexity of multiple projects environment with its challenges in the management aspect and emphasis on the three major domains – inputs, process, and outputs of multiple project environments (Patanakul and Milosevic, 2009).

The review was carried out in several steps: analysis of earlier reviews; literature search in the above mentioned journals; preliminary analysis and collection of key cited articles; and further search for other published work by author. The keywords used were "multi projects", "multiple projects management", "management challenges" and "complexity". During the progress of review, the foundation of the articles was traced. The articles are scanned through to find the most important note of the paper. This was made based on either the author's clear positioning towards certain articles, or the most often cited references. The articles were well-organised according to importance and abstracts were read. When the article met the requirements of the study, the entire article will be read.

Moreover, citation tracking databases such as Scopus and Web of Knowledge are used to search for articles that have been referred to in the studies covered by the literature search and more information also gathered from scholarly books. Despite the limited scope of the literature search, the intention was mainly to support and improve the information and to use existing knowledge on multiple projects management literature as far as possible.

In order to investigate on the challenges in managing the multiple projects environment this paper has reviewed a number of articles from *Project Management Journal, Journal of Manufacturing Technology Management, The Journal of High Technology Management Research, Journal of Engineering, Design and Technology, Journal of Civil Engineering and Management, International Journal of Project Management, IEEE Transactions on Engineering Management, Engineering, Construction and Architectural Management, Construction Management and Economics,* and *AACE International Transactions* for the time period between 2000 - 2012. However, majority of the articles are from the *International Journal of Project Management* (IJPM) because the research published in the IJPM extensively cover a wide range of topics on multiple project environments (Söderlund, 2004).

MPE Challenges

Research into MPEs has been narrowly focused on the challenges within multiple project management. The ambiguous nature of the findings says more about the success factor within researching this complex subject than about the factors influencing the challenges in MPEs (Patanakul and Milosevic, 2009, Dietrich and Lehtonen, 2005, Cooke-Davies, 2002). The studies have supported a conclusion with common sense which assumed the understanding of challenges.

On the other hand, the research evidence to date has relied heavily on complexity in managing multiple projects that cause challenges in management (Caniëls and Bakens, 2011, Aritua et al., 2009). Complexity is not necessarily a new challenge, but an old challenge that is being increasingly recognised and accepted as a key to improving performance and understanding of management (Aritua et al., 2009). Multiple projects management is faced with more challenges than single project management due to the complexity of the environment and organisations related to processes and project lifecycle. The complexity arises from interdependence and uncertainty in management which is the most critical features of context in developing effectiveness in organisational management (Griffin et al., 2007).

Interdependence means that a decision or action by any individual or system may be affected by having different impacts related to other individuals or systems (Mitleton-Kelly, 2003). It is by having many aspects or phases that this decision or action is interrelated. Example of the forms of interdependence are task interdependence in which one job serves as input or output to another job and also interdependence between jobs or roles, team or organisations (Morgeson and Humphrey, 2008). However, there has been little evidence based on the interdependency in management. Further, little research attention has focused on the uncertainty that occurs in managing the multiple projects environment. Uncertainty in management reflects the unpredictability in the inputs, processes and outputs of work systems (Wall et al., 2002). By undertaking activities with lack of specification on comprehensive activity in projects, unfamiliarity of local resources and environment and lack of uniformity will therefore invite an unpredictable environment (de Orue et al., 2009). The organisation will be forced by these factors in meeting project deadlines to achieve higher organisational performance (Laslo and Goldberg, 2008) thus creating challenges for project managers responsible for the overall success of delivering projects (Martyn James et al., 2008).

So far the interaction between the complexity challenge and effectiveness in delivering projects within MPEs within the construction industry is not broadly discussed (Shehu and Akintoye, 2010),even though the acknowledgement on this issue is recognised in practice within the construction industry (Chinowsky et al., 2011, Görög, 2011, Bankvall et al., 2010, Dubois and Gadde, 2002). Conversely, the extensive research on this relationship is dominated by the broader management environment. Researchers (Grant and Parker, 2009, Morgeson and Humphrey, 2008, Griffin et al., 2007, Wall et al., 2002) believe that complexity from interdependence will determine the extent of work roles that are embedded in a broader social system and determine whether an individual is effective by managing their responsibilities individually within an organisation or needs support from broader social context of the organisation. In addition, complexity from uncertainty will determine whether an individual is effective by complying with the requirements of a work role or by adapting to and initiating change. Therefore, since both interdependence and uncertainty are increasing in the management of most organisations, it is especially important to pursue and develop perspectives in managing multiple projects within the construction industry.

Inputs, Processes and Outputs

According to Patanakul and Milosevic (2009) and Hashim and Chileshe (2012), the perspectives on multiple projects environment in the literature offers three related domains: inputs; process; and outputs within the management of multiple projects. By referring to the suggested domains, literature reviews are identified and categorised into the uncertainty and interdependency of challenges that are embedded in managing multiple projects.

Organisational Inputs

The organisational input of the MPE plays an important role to effectively managing the multiple projects. By organisational input, it is referred to the project selection and resource allocation that are influenced by the environmental context of the organisation. Researchers have recognised that project selection and resource allocation are intertwined with the uncertainty in management. Patanakul and Milosevic (2009) have examined the uncertainty in project selection that accentuated linkages with resource allocation in projects. In project selections, organisation should provide essential amount of resources as it is very ineffective to implement too many projects simultaneously without sufficient resources. In addition, the allocation of resources also brings another challenge to project manager (Elonen and Artto, 2003, Fricke and Shenbar, 2000). Lack of adequate resource allocations in every task and difficulty in sustaining resources will influence failure within the multiple projects environment (Patanakul and Milosevic, 2009, Fricke and Shenbar, 2000).

Similarly, Patanakul and Milosevic (2006) also assess the extent to which the selection involves the understanding of project priority as well as to match the ability of project manager with the project assignments. The challenge lies on the assignment of projects to project managers when the number of projects is greater than their ability to manage the projects at a time (John et al., 2000). Effective allocations of projects to the project manager should take into account the priority of a project and match the competencies of project manager with the project requirements recognising personal limitations (Patanakul et al., 2007, Patanakul and Milosevic, 2006, Meredith and Mantel, 2003).

Management Processes

In reviewing current literature, process issues in management are widely discussed and essentially related to uncertainty in the operational level focusing on the project manager. The issues look at the unpredictability in leading the group of projects as being important for effectiveness in management (Patanakul and Milosevic, 2008). Such issues focus on the need for problem solving, information sharing and multitasking typically in conjunction with shifting attention from project to project and responsibility among multiple teams (Patanakul and Milosevic, 2009). In the study by Engwall and Jerbrant (2003) uncertainty in problem solving is formed when the knowledge development is subordinate to solve the short term problem. With regards to information sharing, lack of information delivery within stakeholders will increase in uncertainty (Elonen and Artto, 2003). Besides that, multitasking is demanding in dealing among different issues of different projects that have different goals and characteristics (Patanakul and Milosevic, 2008). In addition, effective communication is about exchanging meaningful information and knowledge among projects and project team to influence thinking and encourage subsequent actions (Lycett et al., 2004). However, good communication is hard to achieve when managing multiple simultaneous projects, thus anticipate challenges in circulating information and knowledge within project processes.

In management processes, interdependency possesses the highest stake. Interdependency represents challenge during the process of implementation and practice stages of construction projects, as the alignment, planning, coordination and execution of the management of single projects in MPEs are carried out with a high level of precision (Shehu and Akintoye, 2010). This stage is critical because the greater the interdependence between processes, the wider the agitation or disturbance of an action by one process on all the other related processes (Mitleton-Kelly, 2003). Elonen and Artto (2003) found that managing multiple internal developments projects is one of the problems. This problem can be seen in terms of the classical concept of project management which includes plan-do-check-action cycles, organisation breakdown structure and work breakdown structures (Hashim and Chileshe, 2012). Hence, any adjustment to schedules or resources which relates to these project management concepts complicates the management of projects due to associated changes among tasks (Patanakul and Milosevic, 2009, Danilovic and Sandkull, 2005, Lycett et al., 2004). In addition, the inter-project interaction becomes an issue when interdependency in MPEs is presented. When one project is under this interaction problems it leads to delay (Fricke and Shenbar, 2000),which affects other projects in return and coordination for align projects to strategy (Milosevic, 2009) as MPEs comprise of projects that are interrelated.

Project Outputs

Typically, a key reason that an organisation implements the MPE is to achieve better efficiency and management of projects. MPE can only be effectively managed in any organisation if there is certainty in project manager's expectation and the project's benefits. The expected output from project managers are meeting time, cost, performance, and satisfying customers and effective use of organisational resources (Patanakul and Milosevic, 2009, John et al., 2000). In the meantime , the project's benefits should be established in the form of potential of project output (Shehu and Akintoye, 2010). Uncertainty about a project's output and benefit can let other projects take priority attention that is not favourable for effectiveness in managing multiple projects.

Discussion

Overall, the emerging perspective on challenges provides important insights into the effectiveness in managing the MPE. This literature review has recognised the attributes of MPEs, which is the management of multiple projects simultaneously (Patanakul and Milosevic, 2009). An example is the assignment of a residential construction project, building construction project and an alteration or improvement of facilities project to one project manager. It also has identified additional attributes of MPE which should be considered to describe the nature of construction industry. One new attribute is the management of multiple projects in multiple locations (Desouza and Evaristo, 2004). One project could be in different region, states, or even country to require benefits from different locations such as close cooperation of professionals or to take advantage of the location. Another attribute is the involvement of multiple organisations that connect stakeholders' actions to the development of projects organisation (Dubois and Gadde, 2002). For example, in one project many organisations are all involved in operations at a construction site which contribute to the resources of various kinds. The organisations also involve in other projects that might cooperate with similar or other organisations. Therefore, each organisation needs to consider different dimensions of co-ordination within the individual project, among different construction projects and inter-firm coordination with other projects.

These attributes should be considered as conditions and moderators of challenges in managing multiple projects. These moderators are uncertainty and interdependency of management that shaped the complexity of the environment (Grant and Parker, 2009) which influence on the organisational input, management processes and output of projects (Hashim and Chileshe, 2012, Patanakul and Milosevic, 2009). Organisational input is the earliest process of the project lifecycles which involves in the initiation or conceptual of the project (Patanakul et al., 2010). The outcomes of uncertainty in the organisational input include project selection in understanding the project priority, matched with the ability of project managers and the project assignments and resource allocation. The management processes is the continuing process throughout the projects which support monitoring and control activities (Project Management Institute, 2008). Uncertainty in the management process takes account of effective communication, leading groups of projects in resolving problems, information sharing and multitasking. While the project output is looking into the overall success of the projects (Patanakul et al., 2010) focus on the project manager's expectations and project benefits.

However, interdependency mainly occurs in the process stage focused on driving the execution of projects (Project Management Institute, 2008) where single projects are manage simultaneously follow by inter-project interactions. Interdependency is important in project management processes for adjusting and linking schedules to match available resources, and removing unnecessary variation in workloads of project managers. Table 1 summarise the discussion on the challenges in the MPE.

Table 1: Summary of the literature related to the challenges in MPEs

MPE / Challenges	Uncertainty	Interdependency
Organisational Input	Project selection *(Patanakul and Milosevic, 2009)* To understand the project priority, match between the ability of project managers and the project assignments Resource allocation *(Elonen and Artto, 2003, Fricke and Shenbar, 2000)*	
Management Processes	Lead group of projects *(Patanakul and Milosevic, 2009, Patanakul and Milosevic, 2008)* -Problem solving *(Engwall and Jerbrant, 2003)* -Information sharing *(Elonen and Artto, 2003)* -Multitasking *(Patanakul and Milosevic, 2008)* Communication *(Lycett et al., 2004)*	Management of single projects *(Shehu and Akintoye, 2010)* -Project Management Process *(Hashim and Chileshe, 2012)* To adjust and link schedules to match available resources, and remove unnecessary variation in workloads of project managers Inter-project interactions *(Milosevic, 2009, Fricke and Shenbar, 2000)*
Project Output	Project manager's expectation *(Patanakul and Milosevic, 2009, John et al., 2000)* Project's benefit *(Shehu and Akintoye, 2010)*	

Contribution

From a theoretical point of view this literature review broadened the project management knowledge in respect to relationships within multiple projects environments and their challenges. The identification of the challenges should be of interest to researchers within risk management in respect to multiple projects environments and this should be recognised as being an essential part of the construction industry. The practical contribution of this study would be through the exploration of challenges in the multiple projects environment that are likely to confront project managers. It should be kept in mind that project management is a core competence and the building of project capabilities thus this exploration is assisting in identifying and mitigating the future risks in managing multiple projects. It also aims to improve the effectiveness and efficiency of project managers by providing findings that serve as a basis for developing strategies within organisational management.

Moreover, in light of globalisation, national and organisational cultures may play an increasingly important role in the MPE development. Within the context of developed countries a comparison guideline should be formulated for the management of multiple projects. National and organisational cultures can play an important role in influencing the challenges in MPEs. However, the lack of in-depth knowledge of how the global environment and the differences in cultures across societies and organisations affect MPEs thus creates challenges that hinder the effectiveness of management. Therefore, it is hoped that this study will inspire other researchers to provide understandings for the development of MPEs not only within the context of developed countries but also for the developing world.

Conclusions

The unpredictability of the current economic climate has directed the development of MPEs. Most studies on MPEs focus on newly development products from the manufacturing industry where the processes are mainly high risk with concurrent processes. However, in the construction industry, the MPE is part of the inherent nature of industry practice, albeit with a lack of research that establishes the understanding of the phenomena. At this stage, project management practice in MPEs has not adopted an explicit way to identify and select the right management style. This study suggests that the understanding of challenges will give rise to adopting the right approach for the right project.

The trends of increasing interdependence and uncertainty in managing projects are emerging which creates challenges for managing the MPE effectively. Ultimately, there is a need for better understanding of challenges in supporting the development of effective management. To achieve this understanding, it may be necessary to consider the various perspectives and challenges in parallel. These advances in MPE research are beginning to answer calls to investigate the challenges and its implications for management. Even though this study applies only to a subset of organisations and industries, challenges in MPEs are relevant to understand and change the experiences and behaviours of managers within the project management discipline.

In summary, what is missing at this point is a comprehensive framework of the challenges and effectiveness in managing the MPE. The need of a management framework is to capture the overall characteristics of the MPE in the construction industry. It will also assist in identifying conditioning variables that influence the relationship and the outcomes they influence, and a core set of mediators and moderators for these relationships. The evidence reviewed above not only aims to provide a platform of progressing into empirical research within developed countries but also as a function of lessons learned to develop a comprehensive study on the development of multiple projects environments within developing nations.

References

Aritua, B., Smith, N. J. & Bower, D. 2009. Construction client multi-projects - A complex adaptive systems perspective, *International Journal of Project Management,* vol. 27, no. 1, pp. 72-79.

Baiden, B. K. & Price, A. D. F. 2011. The effect of integration on project delivery team effectiveness, *International Journal of Project Management,* vol. 29, no. 2, pp. 129-136.

Bankvall, L., Bygballe, L. E., Dubois, A. & Jahre, M. 2010. Interdependence in supply chains and projects in construction, *Supply Chain Management: An International Journal,* vol. 15, no. 5, pp. 385-393.

Blismas, N., G. , Sher, W., D. , Thorpe, A. & Baldwin, A., N. 2004. Factors influencing project delivery within construction clients' multi-project environments, *Engineering, Construction and Architectural Management,* vol. 11, no. 2, pp. 113 - 125.

Blomquist, T. & Müller, R. 2006. Practices, roles, and responsibilities of middle managers in program and portfolio management *Project Management Journal,* vol. 37, no. 1, pp. 52-66.

Caniëls, M. C. J. & Bakens, R. J. J. M. 2011. The effects of project management information systems on decision making in a multi project environment, *International Journal of Project Management,* vol. 30, no. 2, pp. 162-175.

Chinowsky, P., Taylor, J. E. & Di Marco, M. 2011. Project network interdependency alignment: new approach to assessing project effectiveness, *Journal of Management Engineering,* vol. 27, no. 3, pp. 170-178.

Cooke-Davies, T. 2002. The "real" success factors on projects, *International Journal of Project Management,* vol. 20, no. 3, pp. 185-190.

Danilovic, M. & Sandkull, B. 2005. The use of dependence structure matrix and domain mapping matrix in managing uncertainty in multiple project situations, *International Journal of Project Management,* vol. 23, no. 3, pp. 193-203.

de Orue, D. A. O., Taylor, J. E., Chanmeka, A. & Weerasooriya, R. 2009. Robust project network design, *Project Management Journal,* vol. 40, no. 2, pp. 81-93.

Desouza, K. C. & Evaristo, J. R. 2004. Managing knowledge in distributed projects, *Communication of the ACM,* vol. 47, no. 4, pp. 87-91.

Dietrich, P. & Lehtonen, P. 2005. Successful management of strategic intentions through multiple projects – Reflections from empirical study, *International Journal of Project Management,* vol. 23, no. 5, pp. 386-391.

Dooley, L., Lupton, G. & Sullivan, D. O. 2005. Multiple project management: a modern competitive necessity, *Journal of Manufacturing Technology Management,* vol. 16, no. 5/6, p. 466.

Dubois, A. & Gadde, L.-E. 2002. The construction industry as a loosely coupled system: Implications for productivity and innovation, *Construction Management and Economics,* vol. 20, no. 7, pp. 621-631.

Elonen, S. & Artto, K. 2003. Problems in managing internal development projects in multi-project environments, *International Journal of Project Management,* vol. 21, no. 6, pp. 395-402.

Engwall, M. & Jerbrant, A. 2003. The resource allocation syndrome: The prime challenge of multi-project management, *International Journal of Project Management,* vol. 21, no. 6, pp. 403-409.

Evaristo, R. & van Fenema, P. C. 1999. A typology of project management: Emergence and evolution of new forms, *International Journal of Project Management,* vol. 17, no. 5, pp. 275-281.

Fricke, S. E. & Shenbar, A. J. 2000. Managing multiple engineering projects in a manufacturing support environment, *IEEE Transactions on Engineering Management,* vol. 47, no. 2, pp. 258-268.

Gholipour, Y. 2006. 'Multi-Project Resources Procurement in the Construction Industry', in McDermott, P. & Khalfan, M. M. A., eds., *Sustainability and Value Through Construction Procurement,* Salford, United Kingdom, 29 November – 2 December 2006, Salford Centre for Research and Innovation (SCRI), University of Salford, pp. 172-181.

Görög, M. 2011. Translating single project management knowledge to project programs, *Project Management Journal,* vol. 42, no. 2, pp. 17-31.

Grant, A. M. & Parker, S. K. 2009. Redesigning work design theories: The rise of relational and proactive perspectives, *ACADEMY OF MANAGEMENT ANNALS,* vol. 3, no. 1, pp. 317-375.

Griffin, M. A., Neal, A. & Parker, S. K. 2007. A new model of work role performance: Positive behavior in uncertain and interdependent contexts, *Academy of Management Journal,* vol. 50, no. 2, pp. 327-347.

Hashim, N. I. & Chileshe, N. 2012. Major challenges in managing multiple project environments (MPE) in Australia's construction industry, *Journal of Engineering, Design and Technology,* vol. 10, no. 1, pp. 72-92.

John, A. K., Chung-Li, J., Abdallah, S. F. & Wahib, G. J. 2000. Project manager workload--assessment of values and influences, *Project Management Journal,* vol. 31, no. 4, p. 44.

Laslo, Z. & Goldberg, A. I. 2008. Resource allocation under uncertainty in a multi-project matrix environment: Is organizational conflict inevitable?, *International Journal of Project Management,* vol. 26, no. 8, pp. 773-788.

Lycett, M., Rassau, A. & Danson, J. 2004. Programme management: a critical review, *International Journal of Project Management,* vol. 22, no. 4, pp. 289-299.

Martyn James, H., Paul William, F., Martin, S., Carol, K. H. H. & Patrick Sik-Wah, F. 2008. 'The Role of Project Managers in Construction Industry Development', *AACE International Transactions,* DE141-DE149.

Maylor, H., Brady, T., Cooke-Davies, T. & Hodgson, D. 2006. From projectification to programmification, *International Journal of Project Management,* vol. 24, no. 8, pp. 663-674.

Meredith, J. R. & Mantel, S. J. 2003. *Project Management A Managerial Approach,* 5th ed., New Jersey: John Wiley & Sons Inc.

Milosevic, D. Z. 2009. *Program management for improved business results,* New Jersey: Wiley.

Mitleton-Kelly, E. 2003. *Complex systems and evolutionary perspectives on organisations: the application of complexity theory to organisations,* Amsteradm: Pergamon.

Morgeson, F. & Humphrey, S. 2008. Job and team design: Toward a more integrative conceptualization of work design, *Research in Personnel and Human Resources Management,* vol. 27, no. Journal Article, pp. 39-91.

Ngowi, A. 2002. Challenges facing construction industries in developing countries, *Building Research & Information,* vol. 30, no. 3, pp. 149-151.

Ofori, G. 2000. 'Challenges of construction industries in developing countries: Lessons from various countries', in *Proceedings of the 2nd international conference in developing countries: challenges facing the construction industry in developing countries,* Gabarone, Botswana, 15-17 November 2000.

Olford, W. J. 2002. 'Why is multiple project management hard and how can we make it easier?' in Pennypacker, J. S. & Dye, L. D., eds., *Managing multiple projects: planning, scheduling and allocating resources for competitive advantage,* New York: Marcel Dekker, Inc.

Patanakul, P., Iewwongcharoen, B. & Milosevic, D. 2010. An empirical study on the use of project management tools and techniques across project life-cycle and their impact on project success, *Journal of General Management,* vol. 35, no. 3, p. 41.

Patanakul, P. & Milosevic, D. 2006. Assigning new product projects to multiple-project managers: What market leaders do, *The Journal of High Technology Management Research,* vol. 17, no. 1, pp. 53-69.

Patanakul, P. & Milosevic, D. 2008. A competency model for effectiveness in managing multiple projects, *The Journal of High Technology Management Research,* vol. 18, no. 2, pp. 118-131.

Patanakul, P. & Milosevic, D. 2009. The effectiveness in managing a group of multiple projects: Factors of influence and measurement criteria, *International Journal of Project Management,* vol. 27, no. 3, pp. 216-233.

Patanakul, P., Milosevic, D. & Anderson, T. 2007. A decision support model for project manager assignments, *IEEE Transactions on Engineering Management,* vol. 54, no. 3, p. 548.

Pennypacker, J. S. & Dye, L. D. 2002. 'Project portfolio management and managing multiple projects: Two sides of the same coin' in PENNYPACKER, J. S. & Dye, L. D., eds., *Managing multiple projects: planning, scheduling and allocating resources for competitive advantage*, New York: Marcel and Dekker Inc, 1-10.

Project Management Institute 2008. *A guide to the project management body of knowledge (PMBOK® Guide),* Newtown Square, Pennsylvania: Project Management Institute, Inc.

Shehu, Z. & Akintoye, A. 2010. Major challenges to the successful implementation and practice of programme management in the construction environment: A critical analysis, *International Journal of Project Management,* vol. 28, no. 1, pp. 26-39.

Söderlund, J. 2004. On the broadening scope of the research on projects: a review and a model for analysis, *International Journal of Project Management,* vol. 22, no. 8, pp. 655-667.

Turner, J. R. 2009. *The handbook of project-based management: leading strategic change in organizations,* 3rd ed. ed., New York: McGraw-Hill.

Wall, T. D., Cordery, J. L. & Clegg, C. W. 2002. Empowerment, performance, and operational uncertainty: A theoretical integration, *Applied Psychology An International Review,* vol. 51, no. 1, pp. 146-146.

Zavadskas, E. K., Ustinovichius, L. & Stasiulionis, A. 2004. Multicriteria valuation of commercial construction projects for investment purposes, *Journal of Civil Engineering and Management,* vol. 10, no. 2, pp. 151-166.

10

Volatility in Construction: Different Dimensions and Types of Changes

Zhuoyuan Wang, Benson T.H. Lim, and Imriyas Kamardeen
(University of New South Wales, Australia)

Abstract

Effective change management has been touted as the next strategic weapon for competitiveness. The aim of this paper is to investigate the dimensionality of changes in construction. Under this aim, the specific objectives are to: define and classify the types of changes; and operationalize the types of changes. The literature review reveals that changes in construction could categorise along five dimensions: time; need; effect; process; and environment. Each dimension comprises at least two different types of changes. The identified dimensions and types of changes will be tested in the next stage of a research project, and thereafter the empirical findings will attempt to provide industry practitioners in-depth insights into different constituents of change that could affect their business operation.

Keyword: Construction Firms, Changes, Competitiveness

Introduction

Environmental changes in construction are growing at an increasingly rapid pace and are offering consistently greater strategic opportunities with time, while posing significant threats. Change management is thus one of the key approaches for firms to remain viable in a changing business environment (Price and Chahal, 2006; Lewis, 1994). However, managing changes effectively is not an easy task; as changes in construction may vary in types, scale and quantity. Different dimensions and types of changes may bring about varied influences on firms, and require the firms to adopt and configure different responsive strategies in order to remain viable. It is therefore important for firms to appreciate the dimensions and types of changes in construction.

The aim of this paper is to investigate the dimensionality of changes in construction. Under this aim, the specific objectives are to: define and classify the types of changes; and operationalize the types of changes. Ascertaining and operationalizing the types of changes are important because it provides a framework for researchers and practitioners towards appreciating different constituents of changes that could affect business performance, and evaluating the market volatility. The identified dimensions and types of changes will be tested in the next stage of a research project.

Definition of Changes

Ven and Poole (1995) stated that change is an empirical observation in an organizational entity of adaptions in shape, quality or state over time. Boeker (1997) expanded on the definitions of Barr et al. (1992) and Child and Smith (1987), defining change as an adjustment to the environment or an promotion in performance). Change also has been defined as the adoption of a thought or behaviour (a system, process, policy, program, or service etc.) that is new to the adopting organization (Damanpour and Evan, 1984; Daft, 1982; Aiken and Hage, 1971).

Change represents something that people have not expected. Kotter and Cohen (2002) identified change as a reality of people need to deal with new technologies, mergers and acquisitions, restructurings, new strategies, cultural transformation, globalization, and e-business on organizational level. Bridges (2010) defined change as the new site, the new boss, the new team roles, the new policy and other things which are situational, comparing to 'transition' which is a psychological process. In Stagg and Robbins (2008)'s opinion, change in organizations is the factor which brings planning variations, environmental uncertainty, complicate decision-making process, and so on. While Damanpour (1988) defined change on organizational level as either responding to changes in its environment or as a 'pre-emptive' action. According to Raftery and Loosemore (2006), changes could be risks or opportunities to construction industry. In this paper, changes are defined as a situation happened in construction industry which offering consistently greater strategic opportunities with time, while posing significant threats.

Methods

In this research, a literature search was conducted to examine those peer-reviewed articles on changes in business environments and change management in project and organisational contexts. Of these, articles were searched electronically in those databases relating to architectural, engineering, construction and business management (e.g. Science Direct, Emerald, Ecospecifier, Jstor and Sage Journals). Several keywords were used to facilitate literature research. Irrelevant topics such as climate change had been excluded.

The purposes of searching and reviewing literature are: (1) to investigate whether change is an integrative multi-dimensional concept; and (2) what the key dimensions of change are.

Dimensionality of Change

Management literature suggested that change is an integrative multi-dimensional concept rather an independent variable that can be defined and measured in isolation (Johnson and Scholes, 1997; Pritchett and Pound, 1995). However, it appears that there is a shortage of widely accepted and robust capacity to measure change. It follows that the need to develop a generalized set of valid measures for empirical testing of hypothesis about the concept of change.

The review of literature (e.g. Motawa et al. 2007, Johnson and Scholes 1997, and Pritchett and Pound 1995) revealed that the identified dimensions and types of changes could be classified and streamlined into five categories.   . Dimensions and types of changes are discussed below.

Time dimension

Anticipated and emergent changes

Senaratne and Sexton (2011) reported that anticipated changes are planned forward and happen as intended, and, emergent changes emanate spontaneously and are not originally anticipated or intended. The former allows organizations to make more preparation, while the latter should be urgent enough to push the organizations to implement change management no matter whether they are ready or not.

Proactive, reactive and crisis changes

'Proactive', 'reactive', and 'crisis' are the most popular vocabularies to describe changes nowadays. van der Wiele et al. (2001) defined proactive change as the ways that managers use their resources actively to develop society independent of a direct benefit to the firms. Society for Human Resource (2005) believed that reactive change is that kind of change that customers, competitors, shareholders, employees, and other critical stakeholders lease signals to organization suggesting that there are needs to change. Price and Chahal (2006) described crisis change as one kind of change which is driven by the fear of failure and the external environmental factors.

Pre-fixity and post-fixity changes

Senaratne and Sexton (2011) defined pre-fixity changes as the changes that happen in the design development stage, and post-fixity changes are those that occur after the design development stage. This classification is based on one specific stage of the construction project lifecycle. It is have higher occurrence on project change management perspective than on organizational perspective.

Need dimension

Elective and required changes

Ibbs et al. (2001)defined required change as a variation needed to bring the project design in order to comply with a building code, and suggested that it is imperative and should be reviewed and processed differently than an elective change. They added that elective change gives management an option to optimize the original project goals, budget, or schedule.

Discretionary and non-discretionary changes

Egan (2007) described non-discretionary change as dictated change, and, of discretionary change is central to mediocrity.

Effect dimension

The effect perspective is based on the results of changes. Everyone seemed to be terrified of change, actually, not all the changes lead to a negative result. For example, economic change could bring some financial crisis for the construction industry, but the price reduction of construction materials, equipment and labours are undeniable fact. Based on this, changes could be divided into beneficial ones, neutral ones and disruptive changes. These classifications should be considered in the specific situation, the same kinds of change could be beneficial for one company but disruptive for another.

Process dimension

Incremental, punctuated, and continuous changes

The continuous model stays at the highest level, which make organizations to develop the capacity to change itself continuously in a radical manner in order to survive. The punctuated model follows it, which has the primary patterns of activities that are punctuated by relatively short bursts of fundamental change. Incremental model treats changes as being a procedure whereby individual parts of an organization deal incrementally and separately with one problem and one target at a time (Burnes, 1992).

Developmental, transitional, transformational changes

Development changes which doing more of, or better than the current situation. As for transitional change, it dismantles existing ways and formulating new situations and strategies. Transformational changes involve implementing an evolutionary new circumstance, requiring major and continued shifts in organizational strategy and vision(Pritchett and Pound, 1995).

Environmental dimension

Changes could be classified in terms of environment, i.e. internal and external. Internal environmental changes happen due to these elements in the internal of organizations. According to Price and Chahal (2006), the updating of technologies is one example of internal environmental changes. Those external environmental changes come from the external side of organizations, which is more complicated than the internal environment.

Stagg and Robbins (2008) defined the external environment as the forces and institutions outside organizations that potentially have influence on their performance. They also divided the external environment into two parts: the specific environment and the general environment. The specific environment has four important components: suppliers, competitors, public pressure groups, and customers. While global, economic, political/ legal, sociocultural, demographics, and technological are included in the general environment.

Table 1 shows the re-classification of different changes dimensions and types. The differentiating of change in time dimension is based on the time that change had been detected. The change is being detected before or after the time it begins to affect organizations is the key indicator of this dimension. Changes such as proactive change, anticipated change, pre-fixity change are all those changes could be foreseen before it happens, while emergent change, reactive change, crisis change and post-fixity change require organizations to be flexible and vigilant in order to mitigate the impact of those disruptive events on their businesses. Kajewski et al. (2001) recognised the definitions 'proactive change' and 'anticipated change' are overlapped, so they replaced 'proactive change' to 'anticipatory change' when they were researching the cost and difficulty trends of on different changes on time dimension. As for these overlaps, changes on time dimension could be divided into three types: proactive change, reactive change, and crisis change.

Table 1: Re-categorization of change dimensions and types

Dimension	Different types of change		
Time	anticipated	emergent	
	proactive	reactive	crisis
	pre-fixity	post-fixity	
Need	elective	required	
	discretionary	non-discretionary	
Effect	beneficial	neutral	disruptive
Process	incremental	punctuated	continuity
	developmental	transitional	transformational
Environmental	internal	external	

According to Sun et al. (2005), the necessity of the change is the basis of need perspective of change. An elective change is a change that management can choose whether to implement or not, while required change leave the management no choice but to make that change. The words 'discretionary' and 'non-discretionary' in the construction industry is originally used to describe if a project has a choice in implementing or not. It is almost the same meaning with elective or required. So, elective change and required change could answer the represent all kinds of changes in this dimension.

The influence of changes towards organizations is the key criteria in evaluating the effect dimension of change. For organizations, they should launch or utilize beneficial change to increase their competitiveness in the environment, by trying their best to transfer the neutral change into beneficial change, and avoiding or minimizing the negative impacts from disruptive change.

There are six types of changes from two different groups in process dimension. Both groups of changes represent different degrees of changes. Transformational change and continuity change stand for those changes having the highest degree of outcomes - both will achieve radical and evolutionary future states for organizations- while incremental change and developmental change are much like upgrading improvements for organizations. To a certain extent, the two groups are overlapped, but for incremental, punctuated, and continuity changes, they contain one more levels of meaning that whether the change affects the organization in a persistent manner or not, the continuity change having the highest persistent process among these three types of changes. 'Incremental, punctuated, and continuity' are more informative and accurate to describe changes in the process dimension.

According to Stagg and Robbins (2008), the internal and external environments of organizations cannot be evaluated in isolation. The two kinds of environments are inter-linked. An external environmental factor such as an upgrade in labour regulation, for instance, will definitely cause an unpredictable turmoil in organizational structure in the internal environment. But the pattern, process, and strategy of managing a change on organizational structure will never be the same as those on managing a change on regulations and laws. Thus, figuring out which part of environment that the change comes from is not only useful in managing it but necessary.

The Key Features and Connections between Different Types of Changes

Change management is a complicated process of managing an organization, the identification of changes should be delicate, as identifying change is the first step of managing it. It serves as a foundation of the management process, so any mistake could steer the entire change management process to an unknown direction and in turn have a negative consequence on the organization's operation, hence performance.

As listed in the third part of this paper, there are many types of changes, and these attributes of these changes are not stable all the time. It can switch from one kind of change to another one which could be an extreme opposite of the previous one. Identifying the types of changes is not as easy as it looks like. The key features of these dimensions and the differences and connections among these dimensions need to be understood by change management implementers in order to discriminate change types accurately.

From the time dimension, according to Society for Human Resource (2005), there are both advantages and disadvantages relying in each of the timing choices. Timely and carefully-planned implementation is not easy. Lacking the sense of urgency could deter the change management process (i.e. anticipated and proactive change management). However, if the situation is of utmost urgency, management could face many unexpected obstacles in developing systematic plans for effective implementation (i.e. emergent and crisis change management).

The need perspective is based on the demands of organizations. For instance, the required nature of changes is of higher priority than the elective ones in organizational schedule; as the former could exert negative impact on the organisation's business performance if not managed properly and timely. As for the elective nature of changes, they are of less urgency; organizations have more time to develop systematic plans for implementation. To a certain extent, required changes are of similar nature to those of reactive, emergent andnon-discretionary changes on time perspective, while elective changes are similar to those of proactive changes.

Senior and Fleming (2006) described incremental changes as smooth changes that slow, systematic and evolutionary, which share the same feature of proactive changes on the time perspective. Continuous change requires organizations to adopt a proactive attitude to, develop thorough implementation plans. According to Redfern and Christian (2003), developmental change can be either planned or emergent and it proves incrementally over time, while transitional change tends to be episodic and radical. Transformational changes stands at a higher level, signifying the future uncertain state of changes, which cannot be anticipated.

Various changes could have different impacts on organizations, despite the ways that organizations response to them. The attributes of respective changes do have considerable effects on organizations. Figure 1 shows the hypothetical model of organisational change management. In this research, organizations should be seen as complex adaptive and open systems, which inevitably interact with internal and external environmental forces - just like an elastic ball, and the changes are the pressures to this ball. The ways the pressures forcing the ball could explain the processes changes affecting the organizations.

It can be seen from Figure 1 that three dimensions could be set for the sphere: horizon, vertical and the third dimension representing the time dimension, need dimension and effect dimension. The magnitude of the force stands for the process perspective and the location and direction of the force (inside or outside the sphere) could denote the source triggering changes (i.e. from internal or external environment).

For the horizontal axis, which stands for the time perspective, the strength of this direction shows the level of preparedness for the changes. When the strength intensifies, it means that the changes are not of utmost urgency nature, and that organizations have more time to plan and prepare for the relevant changes. The more the organizations are ready for the changes, the more positive effects could effect on the organizations.

The vertical axis presents as the degree of necessity. When the strength intensifies, it signifies that the changes become essential and required for organisations to remain viable.

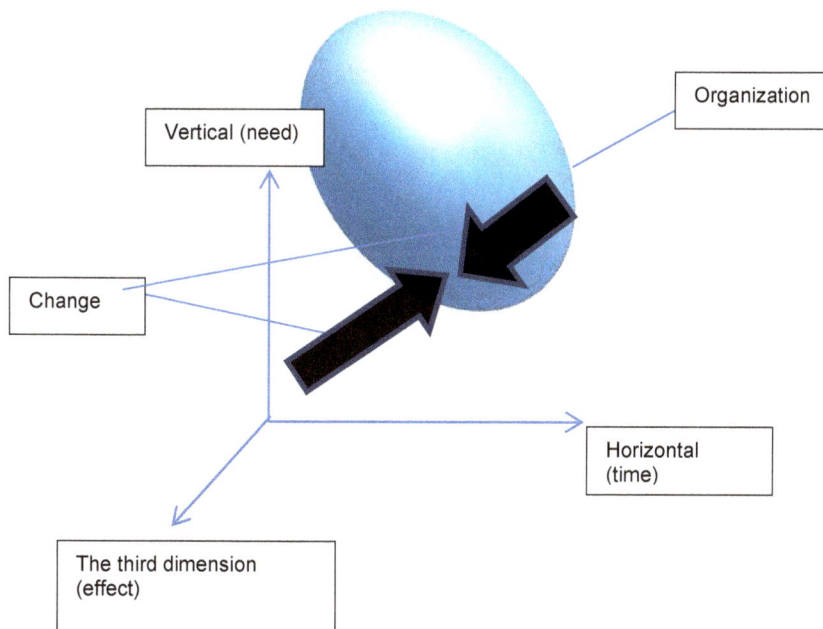

Figure 1: The different dimensions of changes

The effect perspective of changes is demonstrated in the third dimension of the system. Beneficial change stands for the positive position on this direction, while negative position is used for disruptive change. As Raftery and Loosemore (2006) stated, the effect of a change will not remain static, risks and opportunities brought by the change are always stay together. A disruptive change could be beneficial to an organization if an effective management strategy is implemented.

The strength of the forces represents the process perspective of changes. Developmental change has the weakest impact and the highest frequency on the organization, while transformational change has radical influence on it. As for transitional change, it has moderate impact on the organizations. That is because different levels of changes will consume different amount of resources of organizations, and investment for controlling changes and offset the influences of changes could be varied.

The position of changes, as shown in Figure 1, mirror the real-life scenario, it could force organizations to adapt the internal and external changes. For changes from the different environments, even if the other attributes such as time attributes, need attributes, process attributes are all the same, different strategies are required for effective management.

Conclusion

This study investigated the dimensionality of changes in construction. Five dimensions of changes were identified: time, need, effect, environmental and process. Time dimension is related to proactive changes, reactive changes, and crisis changes. The need dimension could be operationalized into elective change and required change, while effect dimension is related to beneficial change, neutral change and disruptive change. As for process dimension, it is dealing with changes of incremental, punctuated, and continuity nature. These dimensions attempt to provide these practitioners a tool to analyse the attributes of changes, evaluate the consequence, and then configure and implement corresponding responsive strategies.

References

Aiken, M. & Hage, J. 1971. The Organic Organization And Innovation. *Sociology,* January, 63-82.

Barr, P., Stimpert, J. L. & Huff, A. 1992. Cognitive Change, Strategic Action, And Organizational Renewal. *Strategic Management Journal,* 13, 15-36.

Boeker, W. 1997. Strategic Change: The Influence Of Managerial Characteristics And Organizational Growth. *The Academy Of Management Journal,* 40, 152-170.

Bridges, P. W. 2010. *Managing Transitions: Making The Most Of Change,* Readhowyouwant.Com, Limited.

Burnes, B. 1992. *Managing Change,* Pitman.

Child, J. & Smith, C. 1987. The Context And Process Of Organizational Transformation - Cadbury Limited In Its Sector. *Journal Of Management Studies,* 24, 565-593.

Daft, R. L. 1982. Bureaucratic Versus Nonbureaucratic Structure And The Process Of Innovationand Change. *Research In The Sociology Of Organizations,* I, 129-166.

Damanpour, F. 1988. Innovation Type, Radicalness, And The Adoption Process. *Communication Research,* October, 545-567.

Damanpour, F. & Evan, W. M. 1984. Organizational Innovation And Performance: The Problem Of "Organizational Lag". *Administrative Science Quarterly,* 29, 392-409.

Egan, G. 2007. *The Skilled Helper: A Problem-Management And Opportunity-Development Approach To Helping,* Thomson Brooks/Cole.

Ibbs, C. W., Wong, C. K. & Kwak, Y. H. 2001. Project Change Management System. *Journal Of Management In Engineering,* 17, 159-165.

Johnson, G. & Scholes, K. 1997. *Exploring Corporate Strategy,* Prentice Hall.

Kajewski, S., Tilley, P., Crawford, J., Remmers, T., Chen, S.-E., Lenard, D., Brewer, G., Gameson, R., Kolomy, R., Martins, R., Sher, W., Weippert, A., Well, G. C. & Haug, M. 2001. Industry Culture: A Need For Change. *Program C : Delivery Management Of Built Assets.* Brisbane: Crc For Construction Innovation.

Kotter, J. P. & Cohen, D. S. 2002. *The Heart Of Change: Real Life Stories Of How People Change Their Organizations,* Boston, Harvard Business School.

Lewis, P. 1994. *The Successful Management Of Redundancy*, Blackwell Publishers.

Motawa, I. A., Anumba, C. J., Lee, S. & Peña-Mora, F. 2007. An Integrated System For Change Management In Construction. *Automation In Construction,* 16, 368-377.

Price, A. D. F. & Chahal, K. 2006. A Strategic Framework For Change Management. *Construction Management And Economics,* 24, 237-251.

Pritchett, P. & Pound, R. 1995. *A Survival Guide To The Stress Of Organizational Change*, Pritchett & Associates.

Raftery, J. & Loosemore, M. 2006. *Risk Management In Projects*, Taylor & Francis.

Redfern, S. & Christian, S. 2003. Achieving Change In Health Care Practice. *Journal Of Evaluation In Clinical Practice,* 9, 225-238.

Senaratne, S. & Sexton, M. 2011. *Managing Change In Construction Projects: A Knowledge-Based Approach*, Wiley.

Senior, B. & Fleming, J. 2006. *Organizational Change,* Harlow, Prentice Hall.

Society For Human Resource, M. 2005. *The Essentials Of Managing Change And Transition,* Boston, Mass., Harvard Business School Press.

Stagg, R. B. & Robbins, S. P. 2008. *Management*, Pearson Education Australia.

Sun, M., Sexton, M., Aouad, G., Fleming, A., Senaratne, S., Anumba, C., Chung, P., El-Hamalawi, A., Motawa, I. & Yeoh, M. L. 2005. Managing Changes In Construction Projects. Engineering And Physical Sciences Research Council.

Van Der Wiele, T., Kok, P., Mckenna, R. & Brown, A. 2001. A Corporate Social Responsibility Audit Within A Quality Management Framework. *Journal Of Business Ethics,* 31, 285-297.

Ven, A. H. V. D. & Poole, M. S. 1995. Explaining Development And Change In Organizations. *The Academy Of Management Review,* 20, 510-540.

11

Quality of Road Construction Projects in Sweden between 1990 and 2010

Abukar Warsame, Han-Suck Song and Hans Lind, (Royal Institute of Technology (KTH), Sweden)

Abstract

It has been reported that productivity developments in construction are slow compared to other sectors. Measuring productivity is however not easy and it might have been underestimated due to deficiencies in the index that is used when the value added in current prices is converted into value added at constant prices. A central problem when deflating nominal prices is that the deflation should only take away price changes for identical products and not price changes related to quality changes and improvements necessitated by new environmental conditions and legislation as well as demand for higher safety. The aim of this project is to help fill this gap by looking at the quality changes over time in road construction. A combination of comparative studies and interviews with practitioners from public and private sectors that are involved in road construction projects were conducted. It was found that there had been both direct quality changes in the roads, primarily to improve safety on the roads and for noise protection, and indirect quality changes related to building in more difficult circumstances because of giving higher weight to other social objectives, especially environmental objectives. As these quality increases are not taken into account when productivity is measured, productivity improvements could be underestimated by around 1% per year.

Keywords: Productivity measurement, Quality improvements, Road construction projects

Introduction

The background for this paper is discussions about productivity in the Swedish construction sector[1]. As reported in Lind & Song (2012) the conventional wisdom is that productivity development in the construction sector has been very slow compared to other sectors. The diagram below, taken from a report by the National Institute of Economic Research (2010), shows that while there was a clear trend towards increasing productivity up to the early 1990s, the figures indicate that there has been no such trend after 1996.

Measuring productivity is however not easy. Lind & Song (2012) paper shows that productivity growth probably is underestimated due to deficiencies in the index that is used when the value added in current prices is converted into value added at constant prices. Swedish Transport Administration annual report (STA, 2010), for example, states that infrastructure condition, quality changes, new legislation and higher safety demands are some of the factors that must be taken into account in order to create a correct picture of how productivity has developed. A central problem when deflating nominal prices is that the deflation should only take away price changes for identical products and not price changes related to quality improvement. The actual quality indicators used are however very crude and therefore the price index used does not distinguish between pure inflation and price increases related to quality changes effectively. This means that reported cost increases per unit can be overestimated as it does not take into account that the higher cost can be related to increasing quality. The question becomes how much quality really has changed during the last 20 years and how large the measuring errors in the productivity figures really are. In order to answer this question, two types of projects in construction sector have been

---
[1] Bygginnovationen fas 1 (2010b), Delrapport Väg. http://www.bygginnovationen.se/sa/node.asp?node=804

analysed more in detail: road construction and residential housing construction. This paper reports the study of road construction in Sweden.

Value added (SEK per hour in vertical axis), seasonally adjusted quarterly data

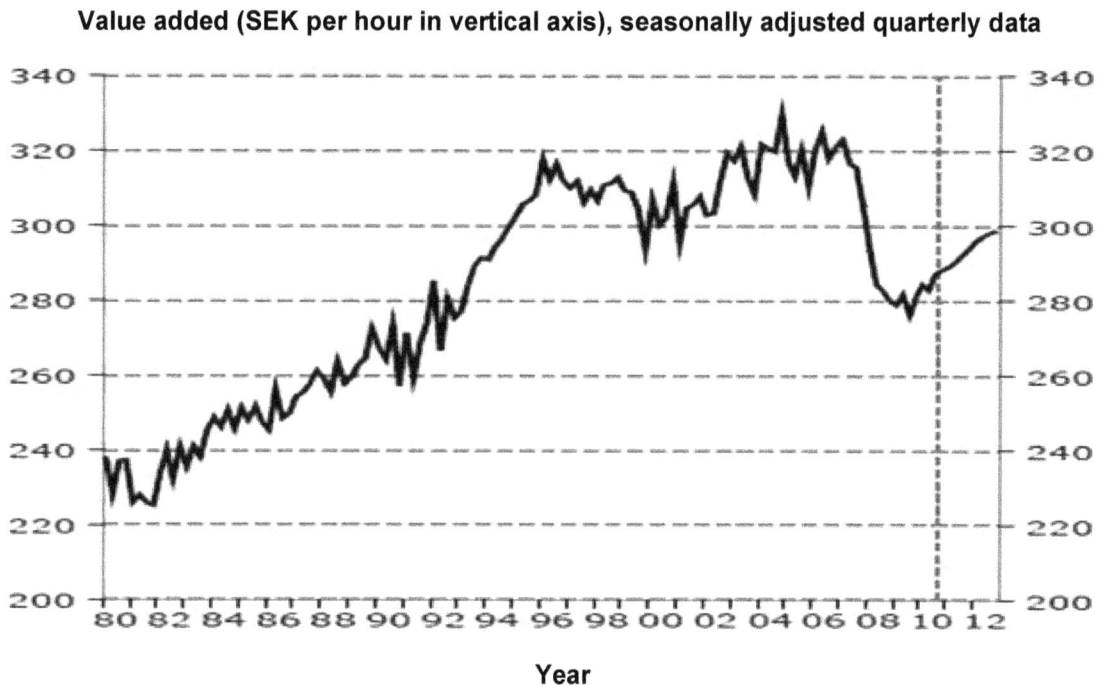

Year

Figure 1 Productivity development in the Swedish Construction sector (value added per hour worked). Source: SCB and National Institute of Economic Research

Cost, time and quality attributes play a big role in determining the level of project performance, and this also means that studies of quality development is of great importance. Ex-ante and ex- post project costs can provide certain indications of budgetary conformance and the development of the level of construction costs over time. For several reasons, this may not be true for the quality development aspect of construction projects. First, infrastructure projects are durable goods that last many decades and thus their performance is strongly dependent on external factors such as weather changes and usage. Secondly, quantification of quality of a construction project is hampered by lack of unique definitions and measurements. Specifications and product or user-based definitions have, for instance been used where characteristics of components of roads are specified (conformance), functionality aspect of product (fit for purpose) or certain level of user satisfaction have been met (Lay, 2009).

The aim of this project is to study the quality changes over time in road construction. Knowledge about quality changes is important both to measure the "true" level of cost increases in the sector and also to measure productivity development in a better way.

Interpretation of Quality

Warsame (2011) gives an overview of quality concepts in the construction sector and it can be noted that a common practice of measuring quality is by the absence of defects. The concept of relative quality is often used to indicate that quality is related to what you had reason to expect given the specifications made in the contract. High relative quality can then occur even in a low-price product, and the other way around. Absolute quality, on the other hand, concerns the specification and the functionality of the product over time in relation to the needs of consumers and society.

Quality changes over time can in principle refer to both relative and absolute quality. In the relative sense, quality improvements would concern primarily fewer defects. A survey conducted by Warsame (2011) found that there were no clear trends in relative quality over time and in this study the focus has therefore been on absolute quality. A newer project might for example include the addition of certain components such as intermediate or noise barrier that were not required in older projects. Safety demands or environmental requirements could necessitate the addition of new components or structures. In this case, incremental quality of projects is reviewed by its presence rather its absence.

It is expected that the quality of construction has changed over time due to technological changes, government regulations and user expectations. Furthermore, there is anecdotal evidence that construction costs have also increased during these periods of time. Bils and Klenow (2000) suggest that the growth rate in unit prices for any good reflects both growth in the average quality of the good and the true rate of price inflation. Inaccurate measures of construction cost increases can, as mentioned above, exacerbate the perceived productivity level of construction industry.

The paper is about quality in the form of changes in the characteristics of the road that affect the "utility" of the road. The focus here is on the quality of the road as an investment object as this is related to the direct cost and therefore the index used for calculation of productivity. In other contexts it might be more interesting to quantify quality improvement over times through more direct performance analysis, e.g. consumer satisfaction and frequency of accidents. The "utility" can concern

- People using the road: surveys are often used to elicit the level of satisfaction of road users. One major concern of deriving quality from this type of utility is the diverse opinion among road users such as private and professional drivers (STA, 2010).
- People building the road: one important quality aspect is measures taken to increase the safety of workers and road users during the construction period.
- Organisations managing the road: road infrastructures have a long life and the main concern for organisations managing them is the cost of operating and maintaining these assets. Quality improvements can take the form of more durable roads with lower life-cycle costs.
- Society at large: Ofori (1992) argues that issues related to environment should be considered as the fourth objective of construction projects. He refers to a report from Economic Commission for Europe that considered measures taken in relation to environmental issues that might increase investment costs, duration of the design and planning processes as well as the construction periods. Another societal aspect that might affect road construction costs is changes in the weight given to archaeological issues.

Method
The basic purpose is to find out the size of cost increases related to changes in quality and the work was carried out in several steps. The first step was to identify the quality changes. This was done through a combination of interviews and comparative case studies. For interviews, after a thorough discussion with different scholars, we concluded that interviews with a very experienced, reliable, and knowledgeable people are more effective and could produce better results. Thus, we have contacted five practitioners (experts) from both public and private sector that we know have had working experience with infrastructure projects.

We have conducted both semi-structured interviews and indirect communication such as e-mails and telephone interview with these practitioners. Our questions were very general and not specific to any projects since our aim of these interviews was to ascertain participants' views on the overall development of road construction projects since early 1990s. The aim was to identify different factors that might have affected the way we design and construct our road construction projects over time. Their candid and holistic view about road construction changes that took place during the last three decades also made it easier to carry out a comparative case studies, described below.

Comparison between similar roads built during in a specific region during different time periods is the other approach. Type of road and climate condition has to be kept constant in order to be able to identify quality changes. In this study the focus is on motorways in the Mälardalen region as it was expected to be rather easy to find such comparable cases. Formally, the method used is described as a multiple case study and the first step is to make a selection of projects. We chose 2 projects that were built around 1990. Secondly we chose 2 projects form the same region as before but were built a decade later, around 2000. Finally, we chose 2 more projects (same region) that were built around 2010. That made it possible to establish if certain component(s) of roads have shown significant specification changes between any two predetermined time periods. Table 1 contains the road segments that were included in the comparative analysis.

Motorway	Location	Open date	Length
E20	Arphus – Härad	1994-06-14	6.7 Km
E20	Härad –Strängnäs-Järsta	2004-10-27	13 Km
E4	E-Länsgräns – Gammelsta	1994-09-19	15.6 Km
E4	Björklinge – Mehedeby Etapp2	2007-10-17	55 km
E18	Örebro –Arboga	2000-10-31	42.8 Km
E18	Sagån – Enköping	2010-10-25	15 Km

Table 1 Motorway projects included

Next we have obtained the descriptions and the bills of quantity of some of the selected projects. However, on the level of project description and the details of the bills of quantities, there are differences between older and newer projects.

In order to identify quality changes a list of components developed by the Swedish Transport Administration cited in Jonsson (2012) was used. The list contains six groups of component types (see table 2). Some of these components in the groups are more likely responsive to the environmental and societal demands while others are solely function of topography and specifications. For instance, changes related to the components of group 2 ("Road construction") and 4 ("Structures") are most likely influenced by topography and structure specifications rather than the utility of end-users. Unlike these two groups, group three ("Road equipment"), five ("Special installations") and six ("Road areas") could be influenced by contextual conditions such as safety requirements and traffic regulations especially when a road is classified as motorway and EU regulations must be followed. Most of major highways cross many urbanized areas, which necessitate more traffic areas and ramps, under and overpasses, as well as foot and cycle path installations to be built. Thus, group five will most likely be influenced by these factors.

Using materials from the different reports, interviews and the case studies it was however possible to find out in general what had happened with the quality of the roads built during the selected period and this is presented in the Results section below.
In the final step, our experts estimated the cost increases that this quality changes have led to and that information was used to estimate how much productivity was underestimated by the lack of information about quality changes.

Groups of component types		
Prerequisite items Road zones Physical planning Archeological surveys Traffic control devices Construction administration	Road constructions Geotechnical constructions Terracing, surface run-offs and road-side areas Superstructures	Road equipments Road markings Guard rails Safety barriers and wildlife fences Road lighting and traffic signals Road information structures
Structures Bridges Large culverts Tunnels Pile foundations Support structures Troughs	Special installations Protection for water catchments Noise barriers Foot and cycle path installations Other special installations	Road areas Rest areas Traffic control and information areas Park-and- ride car parks Buildings

Table 2 Road components Source: Jonsson (2012)

Results

It is not feasible to compare objectively any two projects due to the dissimilarity of design characteristics, terrain and archaeological surveys, design standards for bridges and tunnels as well as other supporting structures. However, the existence of certain components in one project can be identified and cautiously interpreted as an extra improvement of performance of that project compared to another project from an earlier period that is lacking these components. Our rationalized comparison is based on the Jonsson's grouping of road components (see table 2) and it is intended to highlight some noticeable differences of construction components between two segments of same selected road construction project.

Prerequisite Items

The two selected segments of the E 18 motorway construction projects differ in terms of their length (43 and 15 kilometres) and construction costs (878 and 590 million Swedish krona (USD[2] 136 and 91 million respectively). Though economies of scale associated with big projects and price increase of material and labour could be attributed to construction costs differences (21 to 39 million Swedish krona [USD3.3 and 6 million respectively] per kilometres), additional components necessitated by regulation and safety changes as well as environmental requirements could inflate the construction costs of the later project. Increase of width of roads for later projects is also another possible explanation of higher project cost. Based on the presumption that later projects were subjected to fulfil these societal and environmental demands, we were able to identify certain components of recent projects that could be seen as quality improvements.

Preservation of archaeological and cultural heritage sites has big impacts on planning of road construction projects. One of the experts (see below) has mentioned that environmental concerns have affected projects with increased costs.

Road Constructions

It was not possible to ascertain differences between these selected projects in terms of road construction components. However, several practitioners from the industry that we have asked about their opinion about changes of design and specifications of road capacity and

---

[2] Exchange rate Marsh 2014

pavement strength indicated that there have been no significant changes in these aspects in the construction of roads for example the design, thickness of pavements, and specifications of structures during the last 20 years.

Road Equipment

A road safety annual report (IRTAD, 2011) explains how Sweden, over the past twenty years, succeeded in significantly increase the safety of its roads and motorways through the improvement of urban road conditions, the increased use of the "2+1" roads on rural roads since 2000, and more median barriers for motorways.

Despite that the earlier project (Örebro – Arboga) is longer than the later project (Sagån – Enköping) the later project has a higher number of different types of wild life fences and access gates compare to the earlier project. It is not clear if the location of the later project dictated the erection of these components but the influence of environmental regulations cannot be ignored. IRTAD (2011) emphasized that the construction of median barriers for motorways in 2000 partly contributed the increased safety on Swedish roads. Different types of median barriers and railings were also installed at Sagån – Enköping segment that was previously described as Sweden's most dangerous road segment due its number of accidents especially head-on collisions[3.]

Unfortunately we were not able to access information such as the bill of quantity of E-Länsgräns – Gammelsta project. Any observed differences are based on project descriptions from the E-Länsgräns – Gammelsta segment and the bill of quantity from Björklinge – Mehedeby Etapp2 segment. Energy absorbent railing-ends and other crash barrier railings are a noticeable difference.

Structures

Härad –Strängnäs – Järsta segment has more bridges (8 at ~38 million Swedish krona [USD6 million]) than Arphus – Härad segment (only 3 at ~14 million Swedish krona [USD 2.2 million). Topography of the terrain could necessitate construction of these many bridges but another reason could be that Härad-Järsta passes through urban areas and thus requires bridges with under and overpassing structures. It is an open question whether one should see "choosing to build in more difficult circumstances" as a kind of quality change, but it should in any case be controlled for when measuring productivity, as increasing costs might be related to such factors. Here are some excerpts from the interviews that relate to this aspect.

> 'I think we are constructing on more complex sites with more exits, bridges, etc. (exits around cities and not in the middle of the country). In addition, environmental regulations have become tougher and obviously also increased safety and security.'

Special Installations

The Swedish Transport Administration´s objective at the new constructions of roads in the vicinity of existing buildings is to meet good environment guidelines for reducing road traffic noise. Since part of the motorways studied cross urban areas, noise protection barriers were erected along residential areas to a larger extent in the later projects. It is however difficult to control for differences in the specific locations of the roads. Here are some excerpts from the interviews that relate to this aspect.

> 'Intervention through noise barriers (dikes, screens, etc.) and median barriers has affected the cost upwards. Other environmental concerns (Natura 2000) and required

---

[3] http://www.akeritidning.se/svensk-akeritidning/nyheter/svegfors-vill-ha-permanent-overvakning-av-e18-avsnitt

safeguards on other segments of road have also affected the project with increased costs.'

Road Areas

Increase of component group six such as road areas, traffic control and information areas were attributed the safety improvement of Swedish motorways. For instance, rest area with service buildings is included in one of the later projects (Sagån – Enköping at E18) with construction cost of over 3 million Swedish krona [USD 0.5 million]. Other component that is much related to road areas but is not explicitly included in Jonsson's table of road components is vegetation areas.

It is noteworthy that certain costs related to maintenance cost of road projects such as the maintenance of vegetation areas were not found in earlier projects since this type of contracts were not commonly used for older projects. Maintenance of vegetation areas (885 thousand Swedish krona [USD 137.000]) during the warranty period is included in Sagån – Enköping segment. Similarly vegetation and appropriate maintenance of vegetation areas during the warranty period (amount of 1.25 million Swedish krona [USD 195.000]) is one noticeable difference between the Arphus - Härad and the Härad -Strängnäs segments.

Analysis
Direct Quality Changes between 1990 and 2010

The studies presented above indicate the following quality increases during the period under study. Focusing first on the direct changes the following characterized roads built in 2010 but not in 1990:

- middle barrier
- wild life fences
- traffic control areas
- rest and vegetation areas

The next step was then to evaluate the cost increases related to quality changes. As discussed in the method section expert opinions were relied on in this stage. The information above was then presented to experts that we have worked with earlier, followed by the question of how much lower cost would be today if these quality increases had not been made. It was clarified that this was intended to be the same question as asking how much lower costs would be today if we built the 1990-quality instead of the 2010-quality. They were given four intervals to choose between; 0-1%, 1-5%, 5-10% and more than 10%. One expert suggested that if projects were built with same components and standards as 1990 projects, construction costs of current projects will be lower in the range 1-5%. This expert added that cost reductions depend largely on the project's total budget. For small project, the reduction could certainly be between 5-10%. Another expert's response was that construction cost will be lower more than 10% if we would have built current road projects with same quality as in 1990.

Indirect Quality Changes

During the project it was also found that there was another kind of indirect quality increase: *that roads tended to be built in more difficult circumstances in order to satisfy other aims to a higher degree.* One example is choice of location of the road. The simplest way to build might be in an attractive area from a recreational or agricultural purpose and this might have been chosen 20 years ago. But today we choose to avoid these areas and maybe build on more difficult soil conditions. A difference of this type is not covered in the approach chosen above where the focus is on technical changes. In experimental studies that are based on two large road development projects, Heldal (2007) discusses how difficult it is to choose new corridor for a road and not disturb heritage areas. The road planning process especially during the pre-study and inquiry phases will be affected since environmental issues are dealt with at these stages. As Smith et al. (1999) noted, environmental regulations are the outcome of a negotiated process involving different stakeholders and thus it is difficult to measure the impacts. Their study, the only one that has attempted empirically to quantify the impact of environment regulations on construction costs for road projects suggests that regulations increase delays and costs of road construction.

Direct environmental costs on E18 Muurla and Lohja motorway included in our study is estimated to be one per cent of total investment costs of the project. However, there could be other environment related costs that are part of the 4% planning costs since environmental issues are addressed during this phase. Cedermark and von Koch (2000) study, cited in VTI report 530 (2006)[4], indicates that the real cost of mitigation/compensation measures taken was 1 to 9 per cent of the construction cost of the road stretch. An example of this cost increase due to environmental considerations mentioned in that report is E4 segment of 27 KM motorway that was estimated at 750 million Swedish krona [USD 116 million] but alternative designs due to environmental concerns increased the estimated cost by 3 - 4%. In addition, one of the motorway stretches that is included in our comparative case studies (E18 Sågan-Enköpning) is also included in this VTI report with an estimated environmental cost of between 9 to 19 per cent of the total construction cost of that project. The estimated environmental cost for all the 13 case studies in the report was 8-19 per cent.

Total Cost Increases Related to Quality Changes and Implications for Productivity

Table 3 summarize these result indicating a "best guess" of a 13 percentage increase in construction costs related to changes in environmental demand and a best guess of 7 percentage increase related to new and better components in a road project. The total cost increase related to quality increases in the broad sense during the period 1990-2010 would then be roughly 20 per cent.

Factors causing cost increase due to quality changes	Based on literature, case studies and Judgment from expertise	Average
Environment costs	5 to 20 per cent	13%
Cost of added components (railings, middle-barriers, wild-life fences, rest areas, and vegetation areas)	5 to 10 per cent	7%
Estimated average cost increases	10 to 30 per cent	20%

Table 3 Stylized quality changes between 1990 and 2010

As we described in the introduction section, reported cost increases were just seen as increases in prices since the cost increase that is related to quality increases (as figures in Table 3 show) were not taken into consideration. When this price index then is used to deflate the value of the product, it means that productivity changes are underestimated. A neglected quality increase of 20 per cent over a 20-year period means that productivity increases has been underestimate with 0.9 per cent per year. In other words, if this undervaluation of productivity is adjusted, this will reduce the gap between productivity developments in the construction industry and other sectors.

Conclusion

The starting point for this study was weaknesses in the measurement of changes in productivity in the Swedish construction industry, where possible quality improvements were

---

[4] COST 341: Report from European Co-operation in the Field of Scientific and Technical Research
http://www.vti.se/sv/publikationer/pdf/biotopfragmentering-till-foljd-av-transportinfrastrukturen-cost-341-svensk-nationell-kunskapsoversikt.pdf

not taken into account in a satisfactory way. In the presented study the focus was on the building of motorways, but in a parallel study quality changes in housing construction is studied (Borg 2013).

The first result from this study is perhaps that it was very difficult to get relevant data about the quality of road projects. The figures presented should therefore be seen as rough estimates. It was found that there have been both direct quality changes in the roads, e.g. in railings and wild life fences, primarily to improve safety on the roads, and for noise protection. It was also found that there have been indirect quality changes related to building in more difficult circumstances due to giving higher weights to other social objectives, especially environmental objectives. How much this increases cost will of course differ from project to project, but the estimate presented indicates that roads today would be around 20 per cent cheaper if they were built in the same way as 20 years ago. As these quality increases are not taken into account when productivity is measured it indicates that the yearly change in productivity is underestimated by around 1 per cent per year.

A final reservation is of course that the numbers presented are very uncertain due to lack of data, and several researchers in Sweden have pointed out that the Swedish Transport Administration has not made it easy for researcher as their data bases are incomplete and very difficult to use. In general it turned out to be much more difficult than expected to find out the qualities of the different road segments. This was partly because the Swedish Transport Administration was going through reorganisation and partly because data had not been saved in a way that made it easy to find out how the roads were built. Other researchers (Mandell and Nilsson, 2010) have also shown how badly structured is the data collected by the Swedish Transport Administration and how little is done to make evaluations possible.

References

Blis, M. and Klenow, P. (2000) Quantifying quality growth, NBER working paper No. 7695.

Borg, L. (2013) Quality change and implication for productivity development housing construction in Sweden 1990-2010, *Forthcoming working paper*, Building and Real Estate Economics, KTH.

Heldal, I. (2007) 'Supporting participation in planning new roads by using virtual reality system', *Virtual Reality*, 11 (2-3), 145-159.

IRTAD (2011) International Transport Forum, http://www.internationaltransportforum.org/, accessed on 25/5/2013

Jonsson, P. (2012) 'Transport asset management: quality-related accounting, measurements and use in road management' process', Doctoral Thesis in Building and Real Estate Economics, KTH.

Lay, M. (2009) *Handbook of Road Technology*, Spon Press.

Lind, H. and Song, H. (2011) Dålig produktivitetsutveckling i byggindustrin: Ett faktum eller ett mätfel? KTH

Mandell, S. and Nilsson, J.-E. (2010) A Comparison of nit price and fixed price contracts for infrastructure construction projects. No 2010:13, Working Papers, Swedish National Road & Transport Research Institute (VTI).

National Institute of Economic Research (2010) www.konj.se accessed on 28/4/1013

Ofori. G. (1992) 'The environment: the fourth construction project objective?', *Construction Management and Economics*, **10** (5), 369-395

Smith, K., Von Haefen, R, and Zhu, W. (1999) 'Do Environmental Regulations Increase Construction Costs for Federal-Aid Highways? A Statistical Experiment', *Journal of Transportation and Statistics*, **2** (1), 45-60.

STA (2010) Swedish Transportation Administration annual report, www.trafikverket.se accessed on 28/4/2013

Warsame A (2011) 'Performance of Construction Projects: Essays on Supplier Structure, Construction Costs and Quality Improvement', Doctoral Thesis in Building and Real Estate Economics, KTH.

12

Common Risks Affecting Time Overrun in Road Construction Projects in Palestine: Contractors' Perspective

Ibrahim Mahamid (Hail University, Kingdom of Saudi Arabia)

Abstract

The construction sector is one of the key economic sectors and is the main force motivating the Palestinian national economy. However, it suffers from a number of problems that affect time, cost and quality performances. This study aims at identifying the common risks affecting time overrun in road construction projects in the West Bank in Palestine from contractors' viewpoint. 45 factors that might cause delays of road construction projects were defined through a detailed literature review. A questionnaire survey was performed to rank the considered factors in terms of severity and frequency. The analysis of the survey indicated that the top risks affecting time overrun in road construction projects in Palestine are: financial status of the contractors, payment delays by the owner, the political situation and segmentation of the West Bank, poor communication between construction parties, lack of equipment efficiency and high competition in bids.

Keywords: Contractors, Road construction, Time overrun, Risk map, Delay

Introduction

The construction sector is one of the key economic sectors and is the main force motivating the Palestinian national economy. Upon the establishment of the Palestinian National Authority (PNA) and the assumption of its power over the Palestinian territories in 1994, the construction sector has witnessed noticeable expansion and activities. This has resulted in the recovery of the construction contracting profession and subsidiary industries, encouraged the investment of Palestinian expatriates in the local construction sector and contributed to the creation of jobs for thousands of Palestinians. Therefore, the construction sector has occupied a very important position relative to the rest of the economy, attracting investments and creating new jobs (Palestinian Contractors Union (PCU), 2003/1). The Palestinian Central Bureau of Statistics (PCBS) shows that the construction sector contribution to the Palestinian GDP has increased since the PNA establishment to reach 11.1% of the Palestinian GDP in 2010. This is a large proportion that positively affect various other economic, social, educational and vocational sectors and Palestinian institutions (PCBS, 2011/1).

In spite of the high importance of construction sector in Palestine, the industry suffers from a number of problems that affect time, cost and quality performances. Mahamid and Bruland (2012) concluded that all road construction projects implemented in the West Bank during 2004 – 2008 experienced cost overrun. Mahamid et al. (2011) conducted a study to investigate the time delay in road construction projects in the West Bank in Palestine, they found that all projects suffer from time overrun and that 70% of the projects experienced delays between 10% and 30% of the contracted duration. Therefore, attention should be paid to this important sector in order to identify its main challenges and control them. This study aims at identifying the common risks affecting time overruns in road construction projects in the West Bank in Palestine and map the risks from the contractors' point of view. It is hoped that the findings of this study will provide a solid overview and guide efforts to enhance the performance of the construction industry.

Literature Review

Many studies have been conducted to identify the causes of delay in construction projects. Chan et al. (1997) indicated that the five principal causes of delays in Hong Kong construction projects are: poor site management and supervision, unforeseen ground conditions, low speed of decision making involving all project teams, client-initiated variations and necessary variations of works.

In a survey of the West Bank in Palestine, Mahamid (2011) indicated that the most severe factors affecting time delay in road construction projects from the owners' perspective are: poor communication between construction parties, poor resource management, delays in commencement, insufficient inspectors, and rework. Similarly, Al-Najjar (2008) concluded that the most important factors causing time overruns in building construction projects in the Gaza Strip as perceived by contractors were: strikes, Israeli attacks and border closures, lack of materials in the markets, shortage of construction materials at site, delays of material deliveries to site, cash shortages during construction, poor site management, poor economic conditions (currency, inflation rate, etc), shortage of equipment and tools on site, and owner delay in freeing the contractors payments for completed work.

Examining the factors that cause delay in construction projects in Malaysia, Alghbari et al. (2007) tested 31 variables. The main finding of the study was that financial factors are the most common cause of delays in construction projects in Malaysia. Coordination problems are considered the second most important factor causing delays, followed by materials problems.

Al-Momani (2000) investigated causes of delay in 130 public building projects constructed in Jordan during the period of 1990-1997. He concluded that the main causes of delay are related to designer or user changes, weather, site conditions, late deliveries, economic conditions and increases in quantities. Also in Malaysia, Sambasivan and Soon (2007) concluded that the ten most important causes of delays the construction industry were: contractor's improper planning, contractor's poor site management, inadequate contractor experience, inadequate client's finance and payments for completed work, problems with subcontractors, shortage in material, labor shortages, equipment availability and failure, lack of communication between parties, and mistakes during the construction stage.

An interview survey of 450 randomly selected private residential project owners and developers by Koushki et al (2005) found that the main causes of delays in Kuwait were changing orders, owners' financial constraints and owners' lack of experience. Faridi and El-sayegh (2006) studied the delay in construction projects in UAE and concluded that 50% of the construction projects encounter delays and are not completed on time. The most significant causes of construction delays are approval of drawings, inadequate early planning and delays in the owners' decision-making process.

Research Methodology

45 factors that might cause delays of road construction projects were defined through this literature review. The factors were tabulated in a questionnaire form. The questionnaire was divided into two main parts. Part I related to general information about the contractors' companies and the respondents' experience in the construction industry. Part II included the list of the identified causes of delay in road construction projects. These causes were classified into 6 groups according to the sources of delay: project, managerial, consultant, external,

construction items, and financial. For each factor two questions were asked: "what is the frequency of occurrence for this cause?" and "what is the degree of severity of this cause on project delay?" Both frequency of occurrence and severity were categorized according to Table 1.

Scale	Severity	Frequency of occurrence
20%	very low (VL)	very low (VL)
20% - 40%	low (L)	low (L)
40% - 60%	moderate (M)	moderate (M)
60% - 80%	high (H)	high (H)
80% - !00%	very high (VH)	very high (VH)

Table 1 Scale used to identify factor's severity and frequency of occurrence

The average value for responses to each factor is calculated to find out its severity degree and frequency of occurrence. Then, the risk map for factors affecting time overrun in road construction projects in the West Bank was developed based on Figure 1. The map is a 5x5 matrix with severity ranging from VL to VH on the horizontal axis and frequency (with the same range) on the vertical axis (The U.S. Federal Highway Administration Office of International Programs, 2007).

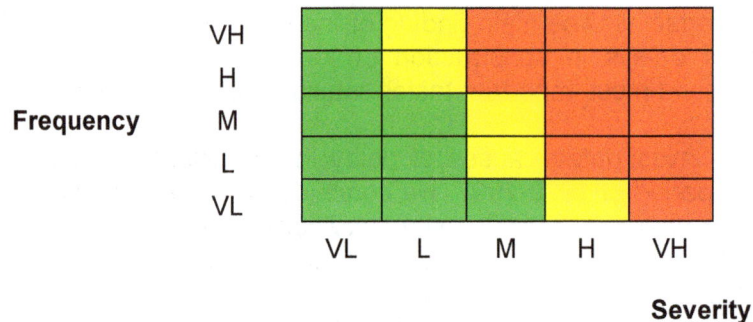

Figure 1 The risk map

The map is classified into three zones:

- Green zone: risks in this zone are low level, and can be ignored.
- Yellow zone: risks in this zone are of moderate importance, and should be controlled.
- Red zone: risks in this zone are of critical importance, and a close attention should be paid.

The questionnaire was sent out to a total of 41 contracting firms asking their contribution in identifying the risk map for the 45 factors in terms of severity and frequency of occurrence. The contractors were randomly selected from available list in PCU including 50 contractors. Only a total of 34 completed questionnaires were returned representing a response rate of 83%. Most of the participating contractors have more than 10 years of experience in road construction.

Results and Discussion
In this study, the risk map for 45 factors that might cause time delays in road construction projects in the West Bank was identified. These factors were classified into 6 groups: project, managerial, consultant, financial, external, and construction items. The risk map for factors under each group is identified in tables 2 – 7 below.

Project Group

Table 2 illustrates the risk map for factors under project group. Six factors are considered under this group. The results indicate that three factors are located in the yellow zone, namely: inconvenient site access, disturbance to public activities, and limited construction area. The other three factors are located in the green zone, they are: poor terrain condition, poor soil drillability, and poor soil suitability.

Factors	Severity	Frequency	Map zone
inconvenient site access	M	M	yellow
disturbance to public activities	M	M	yellow
limited construction area	M	M	yellow
poor terrain condition	L	L	green
poor soil drillability	L	L	green
poor soil suitability	L	L	green

Table 2 Risk map for factor under project group

Managerial Group

Table 3 illustrates the risk map for factors under managerial group. Fifteen factors are considered under this group. The results indicate that two factors are located in the red zone; they are: poor communication between construction parties, and high competition in bids. The other factors under this group are located in the yellow zone.

Factors	Severity	Frequency	Map zone
poor communication between construction parties	H	M	red
delays in decision making	M	M	yellow
unreasonable project time frame	M	M	yellow
internal administrative problems	M	M	yellow
undefined scope of working	M	M	yellow
improper construction method	M	M	yellow
resource management	M	M	yellow
late land hand-over	M	L	yellow
delay in commencement	M	M	yellow
late documentation	M	M	yellow
late submission of nominated materials	M	M	yellow
postponement of project	M	M	yellow
late issuing of approval documents	M	L	yellow
high competition in bids	H	M	red
changes in management ways	M	M	yellow

Table 3 Risk map for factors under managerial group

Consultant Group

Table 4 illustrates the risk map for factors under the consultant group. Eight factors are considered under this group. The results indicate that all factors under this group are located in the yellow zone.

Factors	Severity	Frequency	Map zone
mistakes in design	M	M	yellow
design changes	M	M	yellow
inappropriate design	M	M	yellow
late inspection	M	M	yellow
late approval	M	M	yellow
insufficient inspectors	M	L	yellow
late design works	M	M	yellow
incapable inspectors	M	L	yellow

Table 4 Risk map for factors under consultant group

Financial Group

Table 5 illustrates the risk map for factors under the financial group. Seven factors are considered under this group. The results indicate that 2 factors under this group are located in the red zone, they are: payments delay by the owner, and financial status of the contractor. The results also indicate that 2 factors under this group are located in the yellow zone, and 3 factors in the green zone.

Factors	Severity	Frequency	Map zone
payments delay by the owner	H	H	red
financial status of contractor	H	M	red
financial status of owner	M	M	yellow
exchange rate fluctuation	M	M	yellow
changing of bankers policy for loans	L	L	green
inflation	L	L	green
monopoly	L	L	green

Table 5 Risk map for factors under financial group

External Group

Table 6 illustrates the risk map for factors under external group. Four factors are considered under this group. The results show that the political situation and segmentation of the West Bank are the most important factors under this group and they are located in the red zone. The weather condition and natural disaster are located in the green zone.

Factors	Severity	Frequency	Map zone
political situation	VH	VH	red
segmentation of the West Bank	H	H	red
weather condition	L	M	green
natural disaster	VL	VL	green

Table 6 Risk map for factors under external group

Construction items Group

Table 7 illustrates the risk map for factors under construction items group. Five factors are considered under this group. The results show that one factor in this group is located in the red zone: lack of equipment efficiency. The other 4 factors in this group are located in yellow zone of the risk map.

Factors	Severity	Frequency	Map zone
insufficient labors	M	M	yellow
rework from poor workmanship	M	M	yellow
lack of equipment efficiency	H	M	red
unavailable construction material	M	L	yellow
rework from poor material quality	M	L	yellow

Table 7 Risk map for factors under construction items group

The Critical Factors

Table 8 shows the most critical factors affecting time overruns in road construction projects in the West Bank as well as related groups of factors and comparisons between the results of this study and the studies investigated in the literature review. In this study, 7 factors are found to be critical. These factors are discussed below:

1) Financial status of the contractors: one of the most common problems in construction contracting in Palestine is the policy of awarding the bid to the lowest bidder rather than to the most accurate. The owners award the contracts to lowest bidders, but sometimes the lowest bidder is a less well qualified contractor with low capabilities and resources which leads to poor performance and causes delays in completion of the work. To overcome this problem, the owner may check for the resources and capabilities of the bidders before accepting a bid or awarding the contract, or contracts could be awarded to the bid closest to the client's estimate of the cost and not necessarily to the lowest bidder. This result is supported by Al-Najjar (2008) and Alghbari et al. (2007).

2) Payment delays by the owner: construction works involve high daily expenses that can't be met by the contractors when progress payments by the owners are delayed. This affects the completion of works on time since many of the contracting firms in the West Bank are small with very limited cash reserves. This result is supported by many of the investigated studies (Al-Najjar, 2008; Alghbari et al., 2007; Sambasivan and Soon, 2007, Koushkis et al., 2005)

3) The political situation and 4) segmentations of the West Bank: the political situation in the West Bank is described as unstable because of the conflict between Israel and Palestine. As a result, the West Bank is segmented into many areas that restricts the movement of people, goods and services between these areas. High costs of material, lack of resources, limitations on material import, high level of taxes imposed by Israel, delays and monopolies are some other results of the political situation in the West Bank. All of these factors lead to time overruns in construction projects. This result is supported by Al-Najjar's study of the situation in Gaza (2008).

5) Poor communication between construction parties: this leads to estrangement between the parties and misunderstandings regarding the contract requirements. Thus, this result illustrates the importance of rising awareness among the contracting parties to ensure a culture of team work and to achieve their desires of a less adversarial working climate. This result is supported by several of the investigated studies (Mahamid, 2011; Alghbari et al., 2007; Sambasivan and Soon, 2007).

6) High competition in bids: due to the intense competition between contractors and few projects in relation to the number of contractors in the West Bank, some contractors burn the rates to win bids. In this regard, an accurate estimate of prices may be established by the owner to evaluate properly the tender price before awarding the tender. This may help to avoid awarding contracts to companies offering rates that might affect their ability to execute the project properly and safely. This result was not pointed out by any of the investigated studies.

7) Lack of equipment efficiency: many of the contracting firms in the West Bank are small with inadequate cash flow. Usually, they rent their equipment when required. When there are many construction projects, the equipment are in short supply and are poorly maintained. This leads to failure of the equipment causing projects to be delayed. This result was not pointed out by any of the investigated studies.

Critical factor (this study)	Related group	Severity	Frequency	Map zone	Comparison with the investigated studies
financial status of the contractor	financial	H	M	red	supported by Al-Najjar (2008) and Alghbari et al. (2007)
payments delay by the owner	financial	M	H	red	supported by (Al-Najjar, 2008; Alghbari et al., 2007; Sambasivan and Soon, 2007, Koushkis et al., 2005)
political situation	external	H	H	red	supported by Al-Najjar (2008)
segmentation of the West Bank	external	H	H	red	not concluded by any of the investigated studies
poor communication between construction parties	managerial	H	M	red	supported by (Mahamid, 2011; Alghbari et al., 2007; Sambasivan and Soon, 2007)
high competition in bids	managerial	H	M	red	not concluded by any of the investigated studies
lack of equipment efficiency	Construction items	H	M	red	not concluded by any of the investigated studies

Table 8 Critical factors of this study and comparison with the investigated studies

Conclusion

The aim of this study was to identify the risk map for factors affecting time overrun in road construction projects in the West Bank in Palestine from contractors' viewpoint. The analysis of 45 factors considered in a survey indicates that 8 factors are located in the green zone (low frequency, low severity), 30 factors are in the yellow zone, and 7 factors in the red zone (high frequency, high severity) in the risk map. The most critical factors are: payment delays by the owner, the political situation, the segmentation of the West Bank, the financial status of the contractor, poor communication between the construction parties, lack of equipment efficiency and high competition in bids.

It can be seen that the most critical factors are due to both external and internal issues: the external being high competition, the political situation and segmentation of the West Bank. These factors can't be controlled by the construction parties but the government could reduce these problems to the benefit of all construction parties. The internal issues are: payment delays by owners, financial status of the contractors, poor communication between construction parties and low equipment efficiency. To improve the situation, attention should be paid to these factors, e.g. the owners should pay progress payments on time, the bids should be awarded to the contractors who are financially sound, the construction parties should have more communication and coordination during all project phases, and the managerial skills of the construction parties should be improved by conducting workshops and training courses.

The findings of this study show common critical factors between this study and the investigated studies, namely: financial status of the contractor, payments delay by the owner and poor communication between the construction parties. The contractors also have views on factors that have not been identified previously such as low equipment efficiency, high competitions in bids, and the political situation. This may be the effect of the political situation in the West Bank. The Israeli occupation and boarder control may have negatively affected the performance in construction projects in the West Bank in ways not encountered in other study areas such as: lack of equipment, difficulties in importing materials and new equipment and limitations on the movement of people and goods between the West Bank cities and villages.

References

Alghbari, W., Kadir, M., Salim, A. and Ernawati (2007) 'The significant factors causing delay of building construction projects in Malaysia', *Journal of Engineering, Construction and Architectural Management*, **14** (2), 192-206

Al-Momani, A. (2000) 'Construction delay: a quantitative analysis', *International Journal of Project Management,* **18** (1), 51–59

Al-Najjar, J. (2008) Factors influencing time and cost overruns on construction projects in the Gaza Strip, Islamic University, Gaza

Chan, Daniel W. and Kumaraswamy, Mohan M. (1997) 'A comparative study of causes of time overruns in Hong Kong construction projects', *International Journal of Project Management,* **15** (1), 55-63

Faridi, A. and El-sayegh, S. (2006) 'Significant factors causing delay in the UAE construction industry', *Construction Management and Economics*, **24**, 1167–1176

Koushki, P., Al-Rashid, K., and Kartam, N. (2005) 'Delays and cost increases in the construction of private residential projects in Kuwait', *Construction Management and Economics*, **23** (3), 285-294

Mahamid, I. (2011) 'Risk Matrix for Factors Affecting Time Delay in Road Construction Projects: Owners' Perspective', Engineering, Construction and Architectural Management, **8** (6), 609 – 617

Mahamid, I. and Bruland, A. (2012) 'Cost deviation in road construction projects: The case of Palestine', *Australasian Journal of Construction Economics and Building*, **12** (1), 58 – 71

Mahamid, I., Bruland, A. and Dmaidi, N. (2012) 'Delay causes in road construction projects', *ASCE Journal of Management in Engineering*, **28** (3), 300–310

Palstinian Contractor Union (PCU) Personal interview, March, 2011, Ramallah, The West Bank, Palestine

(PCBS/1) Palestinian Central Bureau of Statistics, The Palestinian economy performance, 2010, Palestine: Ramallah, PCBS, 2011, 43

(PCU/1) Palestinian Contractors Union, Construction Sector Profile. Palestine: West Bank, PCU, 2003

Sambasivan, M. and Soon, Y. (2007) 'Causes and effects of delays in Malaysian construction industry', *International Journal of Project Management*, **25** (5), 517-526

The U.S. Federal Highway Administration Office of International Programs (2007) *Guide to the risk assessment and allocation process in highway construction*, Report, July, 2007, Retrieved from http://international.fhwa.dot.gov/riskassess/images/riskmap.cfm, 10th of April, 2012.

The Use of Cloud Enabled Building Information Models – An Expert Analysis

13

Alan Redmond, (Anglia Ruskin University, UK)

Roger West, (Trinity College Dublin, Ireland)

Abstract

The dependency of today's construction professionals to use singular commercial applications for design possibilities creates the risk of being dictated by the language-tools they use. This unknowingly approach to converting to the constraints of a particular computer application's style, reduces one's association with cutting-edge design as no single computer application can support all of the tasks associated with building-design and production. Interoperability depicts the need to pass data between applications, allowing multiple types of experts and applications to contribute to the work at hand. Cloud computing is a centralized heterogeneous platform that enables different applications to be connected to each other through using remote data servers. However, the possibility of providing an interoperable process based on binding several construction applications through a single repository platform 'cloud computing' required further analysis. The following Delphi questionnaires analysed the exchanging information opportunities of Building Information Modelling (BIM) as the possible solution for the integration of applications on a cloud platform. The survey structure is modelled to; (i) identify the most appropriate applications for advancing interoperability at the early design stage, (ii) detect the most severe barriers of BIM implementation from a business and legal viewpoint, (iii) examine the need for standards to address information exchange between design team, and (iv) explore the use of the most common interfaces for exchanging information. The anticipated findings will assist in identifying a model that will enhance the standardized passing of information between systems at the feasibility design stage of a construction project.

Keywords: Cloud Computing, BIM, Interoperability, Information Communication Technology, Information exchange

Introduction
Research Background

Advancing interoperability between design team applications has been a major challenge for advocates of open standards. Information Communication Technology (ICT) has the capability of streamlining communications between parties at the conceptual design phase to establish an early understanding of the tradeoffs between construction cost and energy efficiency. To a fragmented industry such as construction, the benefits of this service have still to be fully recognized. To the e-Business environment 'cloud computing' is known as the generic term for ICT. It serves as an umbrella term for the provision of services, such as storage, computing power, software development environments and applications, combined with service delivery through the internet to consumers and business.

The building Smart alliance and Open Geospatial Consortium Inc in the U.S. had developed and implemented an Architecture, Engineering, Construction, Owner Operator, Phase 1 Testbed that streamlines communications between parties at the conceptual design phase to establish an early understanding of the tradeoffs between construction cost and energy efficiency (Hecht and Singh 2010). The findings of this Testbed combined with a collaborative Research and Development (R&D) project 'Inpro' Sebastian (2010) co-funded by the European Commission to identify business and legal issues of BIM in construction were used as theoretical propositions underlying this survey.

Research Aims and Objectives

The overall aim of this paper was to establish a model for developing a cloud-based construction service through identifying standardized deliverables, obstacles and opportunities for growth. In order to achieve this aim the following objectives were investigated;

- Identifying the most appropriate applications for advancing interoperability at the early design stage,
- Detecting the most severe barriers of BIM implementation from a business and legal viewpoint,
- Examining the need for standards to address information exchange between design team, and
- Exploring the use of the most common interfaces for exchanging information.

Research Design

This paper presents the results of two Delphi questionnaires. The initial questionnaire undertaken by 16 international experts on construction ICT analyses the expert groups' opinion on the future of ICT in construction based on a cloud service which hosts construction-related applications. The methodology used for the questionnaires included both quantitative and qualitative open and closed-end questions. The attitudinal research focused on subjectively evaluating the opinion or view of the respondent towards a particular topic. The exploratory research was used to diagnose the situation, screen alternatives and discover new ideas. There are two types of experts; those whose expertise is a function of what they know (epistemic expertise), or what they do (performative expertise). An epistemic expertise has the capacity to provide justifications for a range of propositions in a domain while performative expertise is the capacity to perform a skill in accordance to the rules and virtues of a practice (Weinsten 1993).

The following questionnaire compiled the findings of the initial questionnaire and categorised the topics such as, interoperability for BIM software, contractual issues, and information exchange. The original panel of 16 had now reduced to 14. The methodology used for this questionnaire was designed as an extension to the initial questionnaire; for example the initial results for interoperability between three potential BIM applications required further investigation and rethinking. The respondents concern towards vendor reliability and recovering data in the previous questionnaire highlighted the barriers towards a cloud platform but also prompted measures to be investigated for BIM applications relating to contractual issues. Integration of BIM applications on a common database had been signalled out as a major benefit but the issues of successfully exchanging data required at a particular stage needed further research.

Methodology

The Structure of the Initial Survey comprised of the Following Sections:

Business process: The benefits of re-engineering a previous innovative solution with the concept of construction as a manufacturing process were investigated and compared to Kagioglou *et al.* (1999). Kagioglou *et al.* had identified that traditionally ICT had been seen as a driver behind changes in the design and construction process and indeed in many Business Process Re-engineering initiatives.

Cloud computing capabilities: This section Armbrust *et al.* (2009), obstacles to adopting and opportunities for growth of cloud computing, and also investigates Lowe's (2010) review of the five challenges associated with moving backup to the cloud. The final question in this section investigates if cloud computing has advanced from the many mistakes made by the Dot.com bubble (Wohl 2008).

Cloud based business opportunities: This section requests the respondents to refer to their own company when giving an opinion, such as, would cloud be a cost benefit to their firm? The respondent's knowledge is also called into question asking for evidence of expertise on whether cloud benefits are essential for business growth and do Small Medium Size Enterprises (SMEs) have the capability of using such a service.

In relation to the Previous Findings the Structure of the Second Survey comprised of the Following Sections:

Interoperability for BIM software: This section of the questionnaire comprised of questions based on Testbed AECOO-1 and the Inpro projects. The starting question requested the experts' opinion on whether the outlined process for advancing interoperability for BIM software should be focusing on Building Performance Energy Analysis (BPEA), 5D BIM cost estimating software, and information exchange. The second question queried the need for increasing interoperability standards in the BIM marketplace. The 'Inpro project' and Smith (2007) emphasized that by using open standard BIM there was no need to start from scratch as a large amount of systems was already available.

Contractual issues: As this section related to the legal entities associated with BIM, the Inpro project was used. A rating scale of 1 to 5; with 1 being the highest; was used for both questions. The first question listed statements based on the most severe barriers of BIM implementation from the business and legal viewpoints and the second question followed with statements based on the type of contractual terms that should be included in a BIM based project to facilitate open and neutral collaboration processes.

Information Exchange: The information exchange section comprised of two questions formatted to; (i) likert scale, and (ii) rating scale. The first question requested the opinion of the expert by ranking the statements relating to using the industry's most common exchange file mechanism IFC-STEP (Industry foundation Classes – Standard Transfer eXchange for Product Model data) and IFC-XML (eXtensible Markup Language). The second question was structured with two statements taken from Testbed AECOO-1 examining the need for having an open exchange data model. The remaining questions were taken from literature such as Hecht (2008), questioning the use of Sensor Web Enablement (SWE) in relation to BIM, and CISCO and Johnson Controls (2008), analyzing whether BIM Facilities Management would be greatly enhanced by Building Automated Systems (BAS).

Findings and Discussions 'The Future of ICT Through The Use of Cloud Computing'

Business Process
Question 1: The experts were asked if they would agree that developing a cloud collaboration tool based on combining the open Application Performance Interface (API) of accountancy, project management, and BIM applications, would benefit the industry in having a standard supply chain service.

The overwhelming positive response to the question illustrated in Figure 1 showed 50% of the experts agreeing and 29% strongly agreeing. However, after further analysis of this open-ended question, the experts identified areas for concern such as security and the difficulty involved with combining open API's with different applications. The majority of the experts acknowledged that the key to integrated BIM is a common database preferably in the cloud containing information about component parts of building modelled in disparate software programs.

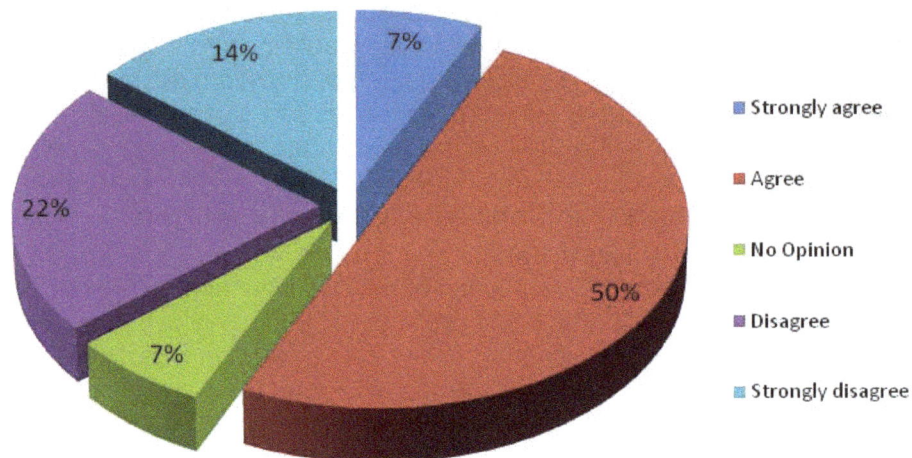

Figure 1 Developing a cloud collaboration tool based on combining open API's of accountancy, project management, and BIM applications

Question 2: A second question asked for the strength of their organisations disagreements with statements made by Kagioglou et al. (1999) regarding the identification of ICT requirements needed to support a process protocol.

The experts had mixed concerns about two statements, namely, (i) the need for a coherent and explicit set of process-related principles to be managed by the whole industry with the intention of changing the strategic management of the common process and (ii) the need for construction operations that form part of a common process controlled by a single integrated team. The problem relating to the first issue can be traced to the fact that companies prefer to manage their own standard procedures until they have to collaborate with the rest of the design team. The second problem refers to the notion of integrated systems being less competitive in comparison to an open standard system. In contrast; the expert panel strongly agreed that the required model should be capable of representing the driver's interest of all stakeholders and be interchangeable allowing interfaces between existing practices.

Other strong indicators identified were the need for a generic and adaptable set of principles, standardised deliverables and a key emphasis on designing and planning to minimise errors during construction. To the question of the construction industry involvement being extended beyond completion a high level of agreement (70%) was evident. The majority of the experts agreed with the process protocols, however, the notion of having the whole industry reviewing the process and controlling the integrated system did receive negative responses.

Cloud Computing Capabilities
Question 3: The third question featured statements relating to the obstacles to and opportunities for growth of cloud computing.

The option of using FedExing Disks (international mail service) to solve the issue of data transfer bottlenecks for large data transfers received an insufficient agreeing response of 43% and a disagreement response of 28%. The no opinion mark of 29% indicated that this should not be the main deterrent and alternative options should be identified. The highest agreement responses by the expert group were allocated to standardizing API's meaning Software as a Service (SaaS) developer could deploy services and data across multiple cloud computing providers. It is important that failure of a single company would not take all copies of customer data with it. The option of scaling storage presented an environmental solution by carefully utilizing resources which could reduce the impact of the data centre on the environment through short-term usage. Scalable storage and data lock-in received a total agreement result of 86% and 85% respectively with no disagreeing responses. Another high level of agreement response was the data confidentially and auditability with its suggested solution of deploying encryption, virtual Local Area Networks (LANs) and networks middle boxes; for example firewalls and packet fillers.

The solution for data confidentially also suggested having geographical storage such as, services located in both the U.S. and Europe to deal with concerns about international law enforcements having the power to search email communications and various records, which received no negative responses. Inventing a debugger that relies on distributed Virtual Machines (VM) also resulted in a high undecided response. Other ideas, such as improving VM support to combat performance and unpredictability, and using multiple cloud providers to prevent Distributed Denial of Service, resulted in an above average agreement response of 57%. The expert group also concluded that the option of pay-for-use licenses did seem attractive. In summary, the solutions to the obstacles were broadly favourable to the group; however the solution for bottlenecks (FedExing) needs more consideration.

Question 4: The experts were asked to rank their opinion on attitudinal statements relating to the major challenges for moving backup to the cloud.

The issue of additional costs increasing because of the lack of knowledge on how backup is perceived and needs meet one's requirements in comparison with one's selected vendor's pricing produced a disagree response of 36% and an undecided response of 14%. This response was the highest disagreement result indicating that it is not a major challenge. In a similar reaction the challenge of backup services outsourced to the cloud with the upstream speeds often capped at very low rates, meaning a cloud-based backup would saturate an upstream connection; received a 57% agreeing mark and disagreeing mark of 29%. A considerable challenge noted by the expert group was security; which is a repeat of the answer for the previous question 3; where deployment of encryption was one preferred solution. However, this statement relates to compliance issues, such as special attention on contractual language, geographical diversity (if your provider offers geographical redundancy in their service) and termination agreements.

In reviewing of challenge of security; the expert group identified a high level agreeing rank of 79%, but this rank was eclipsed by the main concern (that of vendor reliability) which received 100% agreement. This statement presented the issue of negotiating up front about what happens to one's data if a company goes out of business or is acquired. In continuation to the previous challenge; the statement citing the possible solution of working with one's provider to assess their ability and willingness to help one quickly recover from disaster scored an agreement of 86% and an agreement rank of 14%. All of the noted challenges are recognized as a potentially serious issues for customers moving backup to the cloud; however, none more so than vendor reliability and recovery.

Question 5: This question asked the experts for their opinion on the advancements made by cloud computing in contrast to the mistakes made by the Dot.com bubble.

In this question the expert group were requested to rank their opinions on statements relating to the problems of the Dot.com market crash and why cloud computing will not suffer a similar fate. The two most significant corrections were market requirements and a better educated market; both receiving just over 90% agreement mark. The market requirements referred to new cloud applications that attempt to match what the best application in their category offers and then proceed to provide a better interface, better integration with other applications and more web features. A better educated market meant cloud computing would offer access to applications more quickly than traditional decision and implementation processes. This statement also referred to the fact that cloud computing customers do not own the physical infrastructure, instead avoiding capital expenditure by renting usage from a third-party provider.

The market strategy of vendors focusing only on a particular part of a marketplace meant that vendors are not actively focusing on multiple demographics unless they have multiple product and market strategies; this resulted in an agree response of 84%. The issue surrounding stronger business models identifying that cloud vendors plan to monetise their software by either making a charge for each user or each transaction received a modest agreeing response of 67%. This was probably in recognition of the fact that different size enterprises will require different models. The most undecided response of 46% was in relation to better financing; taking into context that venture capitalists have entered in numbers into the market and provided additional development and more sustainable marketing investments. The reason for the expert group's lack of enthusiasm associated with this correction is possibly related to the fact that the western world has not yet recovered from the global recession.

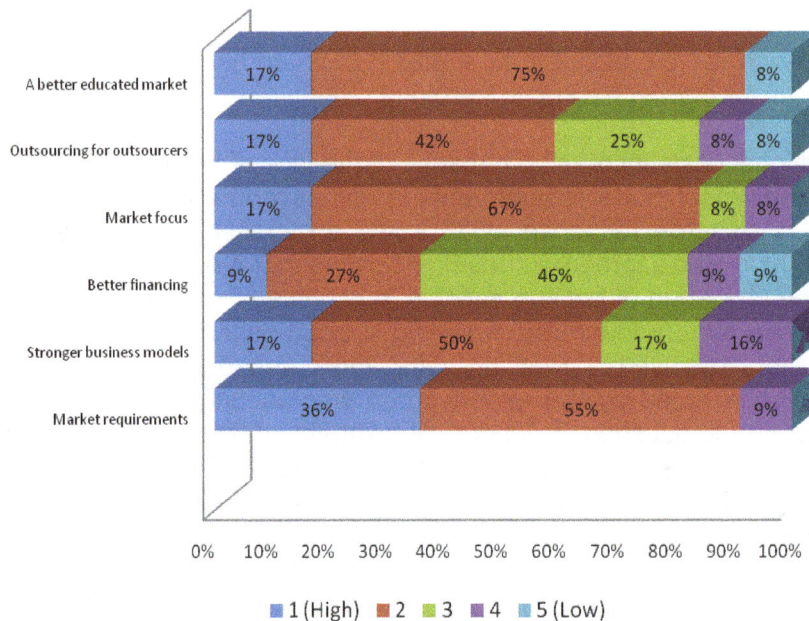

Figure 2 The cloud advancements on the mistakes made by the Dot.com bubble
(Sourced from, Wohl 2008)

The least positive response was outsourcers referring to the idea that vendors now believe it is better to partner for infrastructure than to invest in and run it oneself. The caution shown here was a repeat of the bandwidth issue (if backup services are to be outsourced to the cloud) in the previous question; it was identified as a challenge.

Cloud Based Business Opportunities
Question 6: The experts were asked for their opinion on cloud computing relating to the statements highlighted in Table 1.

In the expert group's opinion there is a lack of knowledge in the construction industry on the various types of construction cloud applications and due to the fragmented nature of the industry; a collaboration tool that provides interoperable software is a necessity. This claim was further enhanced by the group's strong agreement 85% indicating that the future of ICT is a service deployed from a centralized data centre across a network providing access to applications from a central provider. The highest disagreement rating of 38% was related to the notion that the traditional packaged desktop and enterprise applications will soon be made obsolete by web-based, outsourced products and services which is somewhat in contrast to the agreement (77%) response that suggests that cloud computing is an efficient and cost effective outsourcing process that gives company management more time to focus on their business.

Summary of statements supporting cloud computing	Strongly agree	Agree	No Opinion	Disagree	Strongly disagree
The future of ICT is a service deployed from a centralised data centre across a network providing access to the applications from a central provider (cloud computing).	39%	46%	0%	8%	7%
The traditional packaged desktop and enterprise applications will soon be made obsolete by Web-based, outsourced products and services that remove the responsibility for installation, maintenance and upgrades.	23%	31%	0%	38%	8%
The cloud solution generates better opportunities for companies by enabling them to select more ICT priorities from an ever growing menu of applications.	31%	46%	15%	8%	0%
Cloud is an efficient and cost effective outsourcing process that gives a company management the time and opportunity to focus on the core competencies of their business.	23%	54%	15%	8%	0%
Cloud: Pas As You Go (pay for usage rather than for software licenses & hardware infrastructure) is a process that would be of cost benefit to my firm.	23%	46%	23%	8%	0%
Cloud computing present's information risk – but probably not significantly more than in a traditional outsourced environment.	0%	69%	16%	15%	0%
Vendors do not believe that construction SMEs have the capability of using cloud computing.	8%	38%	23%	31%	0%
There is a lack of knowledge in the construction industry on the various types of construction cloud apps.	46%	46%	8%	0%	0%
The downturn in the industry will result in less investment in ICT. This crisis should be used as an opportunity to focus on how to improve things in the long run, cloud computing can act as a major agitator to this concept.	39%	46%	8%	7%	0%
The fragmented nature of the construction industry needs collaboration tools and interoperable software such as, cloud collaborator tool.	61%	31%	0%	0%	8%

Table 1 Summary of statements supporting cloud computing

A similar result was also recorded for the statement that cloud solutions generate better opportunities by enabling enterprises to select more ICT features from an ever-growing menu of applications. The expert group has overwhelmingly stated that cloud is the future of ICT but they are still reluctant to predict that this is the end for traditional packaged desktop and enterprise applications. The pay-as-you-go payment option only received a modest 69% approval; however the response to this question may have also been affected by also asking the respondent would they themselves implement it. In contrast to questions relating to the risk of security, the view that cloud computing presents information risk, but probably not significantly more than in a traditional outsourced environment, indicates that the group does acknowledge cloud's credibility. Redmond *et al.* (2010) identified through a study of the barriers for adoption of cloud computing that vendors do not necessarily believe that construction SMEs have the capability of using cloud computing. The expert group's opinion on this matter resulted in a mixed outcome of 46% in favour, 23% undecided, and 31% disagreeing.

Question 7: The experts were asked to rank perceived benefits of cloud computing for the construction industry.

The eight proposed perceived benefits of cloud computing from Ramanujam's (2007) key points as to why Cloud/On-Demand would be a smart choice for companies. In analysing the expert group's responses again disaster recovery was evidently a concern, with the experts indicating a disagreement response of 23%. The highest disagreement response of 25% was directed towards having the ability to manage a premise-based facility so attention can be redirected towards the customer. The highest agreeing response of 92% highlighted the benefit of allowing one to pay-as-you-go, pay for usage rather than for software licenses and hardware infrastructure. This was in contrast to the previous question (summary of cloud computing) where the respondents only delivered a 69% confidence mark. Both managing a premise facility and frequent updates had the highest undecided percentage rank of 25%. The notion of having access to the best of breed technology did; however; result in a positive 75% whereas managing a premise facility represented the most negative responses of all the benefits.

Findings and Discussions 'Evaluating a Cloud Integrated Model through BIM'

Advancing interoperability for BIM software
Question 8: This question asked the experts to identify whether the following three processes were the most favourable option for advancing interoperability for BIM software (i) BPEA, (ii) Quantity Takeoffs for Cost Estimation, and (iii) Request for information.

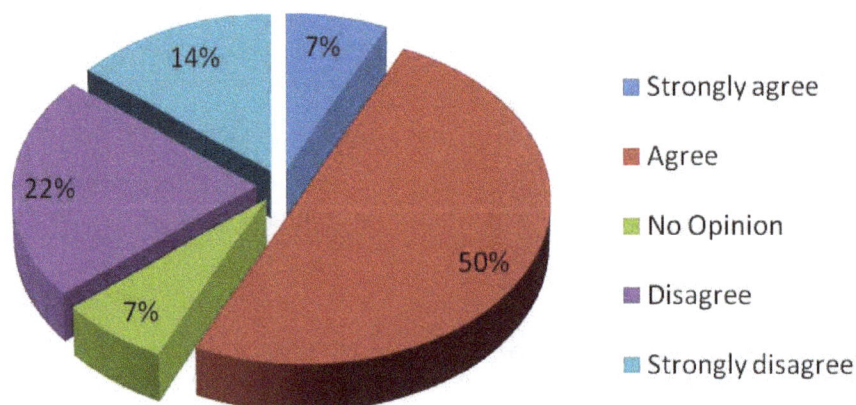

Figure 3 Advancing interoperability for BIM software

The results indicated that 57%, representing 8 respondents, felt that the three most favorable services for advancing interoperability for BIM software are (i) BPEA, (ii) Quantity Takeoffs for Cost Estimation, and (iii) Request for information (RFI). However, 36% disagreed and one individual had no opinion. The question itself tested the idea that the most beneficial stage to advance interoperability is at the conceptual stage and that the three main business areas that are most likely to require interoperability are as previously stated. The open-end answers recognized several different approaches to advancing interoperability such as one respondent's view that data transparency and quality, spatial co-ordination, understanding of data in a spatial context, and management of the supply chain data are the main business processes. Another respondent identified RFI workflows, quantities and estimating, and quantities by location for scheduling. There was also a respondent who correctly pointed out that building performance is not only about energy, but it is also about comfort and future services provided by buildings. In summarizing the result of this question over half of the respondents agreed that the most favourable process for advancing interoperability for BIM software is (i) BPEA, (ii) Quantity Takeoffs for Cost Estimation, and (iii) Request for information.

Question 9: The experts were questioned on their opinion of increasing interoperability standards in the BIM marketplace as highlighted in Table 2.

Increasing interoperability standards	Strongly Agree	Agree	No Opinion	Disagree	Strongly Disagree
The market is increasingly demanding that open standards be more broadly applied to BIM.	14%	79%	0%	7%	0%
Viable software interoperability requires the acceptance of an open data model.	22%	50%	14%	14%	0%
Within a design project, there is little need to share all aspects of the design between project participants.	21%	36%	0%	43%	0%
Multidisciplinary project teams that share tools and information achieve better results than using traditional applications.	36%	57%	7%	0%	0%
With open-sourced BIM designers can plug into an existing variety of typologies, systems and subsystems.	7%	36%	36%	21%	0%

Table 2 Increasing interoperability standards in the BIM marketplace

The notion that the market is increasingly demanding that open standards are more broadly applied to BIM technologies; so that each partner in a project can comfortably adapt their internal processes; received a majority positive indication of 93%. Only one respondent disagreed which clearly identifies that the way forward for interoperability for BIM is to engage in open standards. The second statement; relating to viable software interoperability in the capital facilities industry requiring the acceptance of an open data model and the use of service interfaces contained within provider's software; obtained a positive 72% and negative 14% with another two respondents indicating no opinion. This 72% can be seen to support the National Building Information Modeling standards view that an open data model would provide an industry-wide means of communication enabling every software application used across the lifecycle to become interoperable. The Testbed AECOO-1 maintained that within a design project, there is little need to share all aspects of the design between project participants, and what is relevant is to exchange elements of design between the lead architecture firms or lead general contractor and subcontractor with specific areas such as lighting, energy usage, building cost and Heating Ventilation and Air Conditioning (HVAC). The expert panel projected a mixed response to this statement with only 57% agreeing. The 43% of the expert panel that did not entirely agree with the Testbed AECOO-1 model was because some of the experts are inclined to believe that all information should be shared no matter what process stage the project is reviewing. The next statement was designed to clarify the need to have interoperable applications shifting away from legacy systems. The response from the expert panel clearly agreed with this concept delivering a positive response of 93% and only 7% having no opinion. The final statement referred to the idea that by using open-standard BIM, designers do not need to start from scratch as a large variety of building typologies systems and subsystems are available as the basis of their design.

This enables buildings with high architectural quality to be designed, produced and delivered according to systematic procedures which allow effective control and value optimization for the clients and end users. Only 43% of the expert panel agreed with this concept, 36% had no opinion and 21% disagreed. The open standard content was meant to represent a model server and open communication platform for information sharing. It is possible that the expert group confused this with Open Source Software (OSS) where co-operation is promoted between the user and owner of a software product by removing obstacles imposed by the owner, such as copyright law. The overall conclusion of this section depicts that there is a need to share information through open standards with an industry demand for applications to become more interoperable.

Contractual Issues

Question 10: The experts were asked to rank in order the most severe barriers to BIM implementation from the business and legal viewpoints as indicated in Table 3.

The barriers were categorized into five main issues and structured in a rating scale format. The first barrier signified that there is a lack of immediate benefits of BIM for the stakeholders. This produced a response of 50% disagreeing and 36% agreeing indicating that the expert panel partially sympathizes with the stakeholders need for Immediate ROI. However, the 50% level of disagreement demonstrates that there are immediate benefits to BIM possibly referring to its ability to identify early cost savings. The next barrier highlighted the issue of changing roles, responsibilities and payment arrangements resulting in 50% agreeing, 21% disagreeing, with no opinion at 29%. The Inpro project claims that there is a lack of clarity over the changing roles and responsibilities; for example is the architect still the lead designer in the integrated design and engineering? Who is in charge of the total quality of the design? Who assures that all interface problems (clashes) are solved and that the model is fully secured? These are just some of the issues and the results of the expert panel showed a 50% acknowledgement of this barrier and 29% unsure which demonstrates that this is an issue that needs to be resolved. The barrier associated with the uncertainty of the legal status and intellectual property rights of the model generated a high (79%) agreement with this statement of which 22% of the panel ranked it as a number 1 (the highest barrier) and only 14% disagreed. The major issue relating to this barrier is to what extent anyone can claim ownership of the intellectual property; if the model is deemed to be collaborative work, then ownership may not be vested in a single party.

Barriers of BIM Implementation	1 (High)	2	3	4	5 (Low)
Lack of immediate benefits	7%	29%	14%	36%	14%
The changing roles	14%	36%	29%	14%	7%
Uncertainty of the legal status	22%	57%	7%	7%	7%
Inadequacy of existing frameworks	23%	39%	23%	15%	0%
Lack of consent on protection of information	7%	14%	43%	22%	14%

Table 3 Barriers of BIM implementation

The following barrier; concerning the inadequacy of the existing contractual frameworks, including the agreements on liability and risk locations; presented a response of 62% agreeing and a no opinion of 23%. There are major concerns with who is liable for information in the digital model and how the users are protected and this may be the reason for the 62% of the expert panel agreeing with this barrier. The final barrier referred to the lack of consensus on the protection of information in conversion and interoperability and against loss and misuse of data.

The response received a mixed reaction from the panel with 43% identifying no opinion, 36% disagreeing, and 21% agreeing. The barrier itself is related to the notion that there is a requirement within the industry for an agreement on the standard of care and possible conflict resolution on data management as an integral part of the contract. The results of the survey are inconclusive possibly because there are already standards and agreements available for use of data management; however they are country-specific. The results of the 5 statements emphasized the major barriers to implementing BIM with the structure of a single model created by many disciplines as the main problem due to claim of ownership, who is liable, who is in charge of the total design, whether it should be an integral part of a contract, and can the stakeholders benefit of such a model.

In response to the open-ended question requesting further barriers to be identified; the expert panel views varied. One expert claimed that there is a lack of understanding of how to use BIM and lean business methods particularly in a collaborative business arrangement; this view was supported by another expert who also considered the lack of understanding of how to effectively use BIM in a team environment as a major barrier.

Question 11: This question asked the experts to rank contractual terms that should be included in a BIM based project to facilitate an open and neutral collaboration process, as illustrated in Figure 4.

The previous question list of barriers analyzed the problems associated with implementing BIM. This question focuses on rectifying that situation and identifying the contract clauses needed. The question was formatted to rating scales 1 – 5 (1 being the highest). The first statement highlighted the contractual issue of agreeing on modeling protocols, sharing and integration of open technology and then proposing a solution: endorsing internationally accepted open standards. The expert panel rated this option 58% in favour and 24% against.

The ability to have clauses relating to the workflows, level of authorization, and access rights in a BIM based decision-making virtual project received a response of only 50% in favor and a low 14% against. The highest no opinion of all the statements (36%) was in reference to this statement. This response was very much in line with question 3 relating to the level of clarity over the changing roles and responsibilities, where the results of both statements indicate a high no opinion and an average of 50% in favour. The concept of including a clause for the intellectual property of the foreground and background information and knowledge provided an average 62% in favour.

The Inpro project had perceived a possible solution for the legal status of such a model by enabling the model to serve as a contract document that is used between contractual parties, but is not to be submitted to permit-issuing agencies. This received a favorable response of 61%, 23% no opinion and 16% not in favour. The Inpro project identified that if this is not the case the model may become a document which provides the visualization of the design intent from contract documents. This may also be the reason why the results reflecting the previous solutions have such a mixed response. The final statement emphasizes that depending on the selected contract form and procurement method, particular contract terms should be considered as additional clauses to the contract. The clauses identified were establishment of partnering and the legal entity of the enterprise, format roles and responsibilities, agreement on payment features, and dispute resolution using BIM. The survey supported these clauses with 79% of the expert panel agreeing and only 14% not. In summarizing the key issues identified, one expert stated 'integrated agreements only work if the team members trust each other, trusted business relationships emerge over time, it is naive to think we can "catch the magic in a bottle" via a contract'.

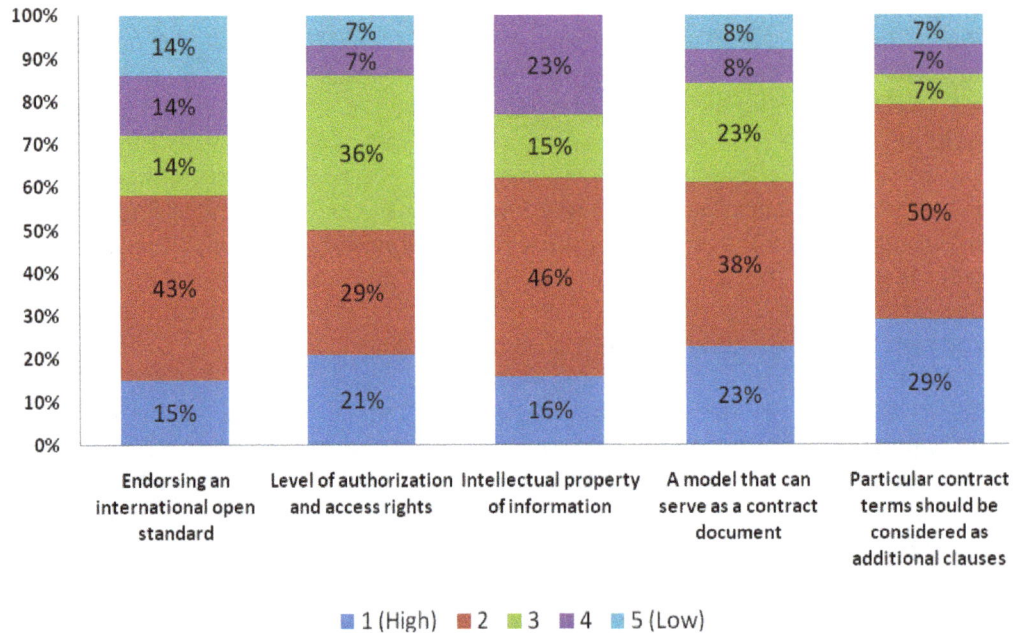

Figure 4 Contractual terms to facilitate open and neutral collaboration

Information Exchange
Question 12: The experts were asked to indicate their opinion on the statements relating to IFC and XML, as highlighted in Figure 5.

This question allows analysis of statements based on whether the construction industry will pass files via STEP or XML. The first statement perceives the computer language EXPRESS; which is one of the main products of ISO-STEP and used to represent conceptual or abstracted objects, materials, geometry, assemblies, process and relations as an foreign format for providers to maintain and stresses that it is not presently in their code product offerings and that IFCs will continue to be marginalized. The results showed that there were an equal number of experts who agreed to those who probably did not understand the statement with comments such as, 'I'm not sure what EXPRESS is but it sounds bad' (36%), while 28% disagreed. The following statement investigated the issue of using an EXPRESS language to pass information in a web service, and referencing it as a poor fit with insufficient mainstream market adoption. This statement received a high no opinion of 57%, 28% disagreeing and only 15% agreeing. The notion that the industry has already moved towards the exclusive use of XML standards with encodings such as Open Building Information eXchange made for web services integration to BIM software resulted in a no opinion and disagreement of 36% with only 29% agreeing.

In review of the previous statements; the final statement summarized that XML is designed to work with web services and there is already available software standards to facilitate the adoption of existing AEC-based XML encoding and schema. The issue as to whether EXPRESS creates an extra cost barrier received a high no opinion from the expert panel (43%). However, 33% of the expert panel did agree in comparison to 14% disagreeing.

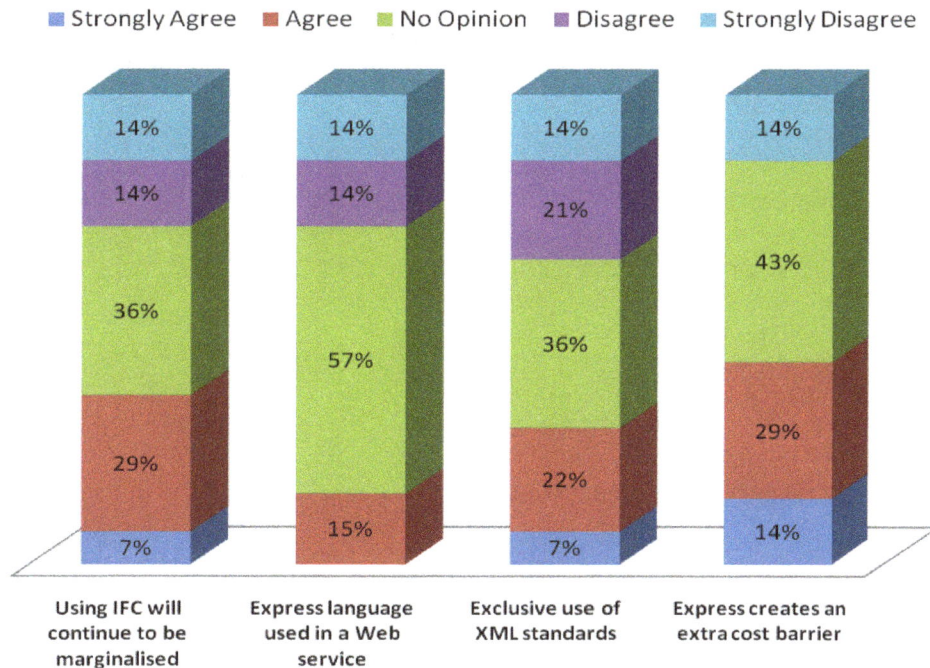

Figure 5 shows the respondents preference between IFC – XML

The open-end question provided mixed comments from the experts, with one expert openly stating that they were unsure of what EXPRESS means. For those who did, the response varied from stating that XML is fine for a quick fix but is problematic for the long haul, to the identification from another expert that they are trying to incorporate Associated General Contracting (agc) XML (a set of XML schemas designed to automate and streamline the exchange of information) and IFC-compatibility. In summarizing the comments of one expert was 'a robust model such as EXPRESS needs to underpin a complex environment such as this, and there needs to be debates as to how data and process tools are implemented as web services, but you cannot escape the need to finish the modeling and design'.

Question 13: This question asked the experts in their opinion to rank statements based on information exchange requirements, as shown in Table 4.

This question posed a series of statements relating to information exchange and the concept of using semantic tagging, sensor web enablement and Building Automated Systems. The initial statement targeted the industry's requirements for software interoperability through exchange definitions, adoption of an open exchange data model and a common interface to the exchange data model for use by any participating application. The results were overwhelming in favor of this concept with only 7% both disagreeing and no opinion. The following statement reviewed the concept of using MVDs and IDMs for incorporation in specifications to be implemented in software. The majority of the experts agreed with this concept (76%). The notion of using semantic tagging in assisting the overall schema for building information in identifying (i) energy efficiency, (ii) manufacturer name, (iii) serial number, and (iv) warranty received 50% in favour but a split between no opinion (29%) and disagreeing (21%) equaled the positive response. In identifying if sensor web enablement (a type of sensor network on geographic information system that is especially suited for environmental monitoring) should be incorporated into a BIM model to optimize energy usage, the expert panel gave a negative indication with 36% of the panel having no opinion and 43% disagreeing, while only 21% were in favour. The final statement is similar in context to the previous statement (analyzing a system for facilities management) in reviewing whether a BAS for importing HVAC after hours and utility meter readings into accounting systems and automatically generating tenant bills that would greatly enhance a BIM Facilities Management system. The expert panel was more in favor of this concept with 46% agreeing, however 23% had no opinion and 31% disagreed, illustrating that neither of these FM systems indicated potential successful adoption.

Information Exchange	1 (High)	2	3	4	5 (Low)
Adoption of an open exchange data model	36%	50%	7%	0%	7%
Incorporating IDM and MVD into specifications	29%	43%	14%	0%	14%
Semantic tagging assist overall schema for building information	0%	50%	29%	14%	7%
Sensor Web Enablement incorporated into a BIM	14%	7%	36%	36%	7%
BAS would greatly enhance a BIM FM	15%	31%	23%	8%	23%

Table 4 Information exchange requirements

Conclusions

The majority of respondents viewed cloud computing as a positive form of physical infrastructure that would increase efficiency and productivity. The notion of using an integrated BIM process through a cloud service was registered as a key benefit to component parts of the building modeled in disparate software programs. The 3 main core sections of the initial questionnaire, business process, cloud computing capabilities, and cloud-based business opportunities all provided evidence that a service based on cloud computing and standardized deliverables would enhance greater market opportunities for the construction industry. Cloud-based as-built-BIM was acknowledged as a service that would increase business decisions. However, whether applications such as, accountancy and project management should also feature as the main drivers failed to encourage a confident conclusion.

In further analyzing the main 3 BIM applications to be tested for advancing interoperability at the early design stage, BPEA, 5D, and request for information were deemed the most favorable. The issue of using propriety file based exchange mechanisms between BIM applications was viewed as a negative approach in comparison to the market demands for open standards between multidisciplinary project teams. A centralized web hosted database was recognized as the main platform for enhancing standardized passing of information between systems. However, ownership and who is in charge of the model are significant barriers against implementations.

The process of using IFC-STEP in comparison to IFC-XML favored XML because of its web services Integration ability with BIM. However, on further investigating EXPRESS language the majority of the respondents were unsure of its meaning. In respect to semantic tagging, SWE and BAS for incorporation into BIM the majority of responses relating to a high no opinion reflected a lack of knowledge on the topic. Overall, the respondents did acknowledge the potential benefits of a service model based on 'Cloud BIM' for analyzing energy performance of buildings through the use of 5D estimating. The results of this research has determined that the market is increasingly demanding that open standards are to be applied to BIM and that be having multidisciplinary project teams that work together with data sharing tools and common information models can exchange information faster than standard legacy systems possibly through the use of web services.

References

Armbrust, M., Fox, A., Griffith, R., Joseph, A.D., Katz, R.H., Konwinski, A., Lee, G., Patterson, D.A, Rabkin, A., Stocia, I. and Zaharia, M. (2009) 'Above the Clouds.'A Berkeley View of Cloud Computing, Electrical Engineering and Computer Sciences University of California at Berkeley.

Cisco and Jonhson Controls. (2008) 'Building Automation System over IP (BAS/IP)', Design and Implementation Guide, www.cisco.com/go/validateddesigns, Cisco Validated Design. V8.1

Hecht, L. (2008) 'A Sustainable Building Industry Requires Service-Based BIM', V1 Magazine, Promoting Spatial Design for a Sustainable Tomorrow, http://www.vector1media.com/

Hecht, L. Jr and Singh, R. (2010) 'Summary of the Architecture, Engineering, Construction, Owner Operator Phase 1 (AECOO-1) Joint Testbed', *buildingSMART alliance (bSa) and The Open Geospatial Consortium, Inc. (OGC),* http://portal.opengeospatial.org, Discussion Document

Kagioglou, M., Cooper, R. and Aouad, G. (1999) 'Re-Engineering the UK Construction Industry: The Process Protocol', *Proceedings of the Second International Conference on Construction Process Re-Engineering* (CPR-99), 12-13 July, University of New south Wales, Sydney Australia

Lowe, S (2010) 'Five Challenges for moving backup to the Cloud', *TechRepublic,* http://www.techrepublic.com/blog/datacenter/five-challenges-for-moving-back-up-to-the-cloud/2618, 23/06/2010

Ramanujam, B. (2007) 'Moving SaaS/On-Demand from Dream to Successful Reality', *A Practical Solution to Real-World Problems with Contact Centre Adoption SaaS/On Demand,* http://www.cypcorp.com/leadership/ index.php

Redmond, A., Hore, A.V., West, R.P. and Alshawi, M. (2010) 'Building Support for Cloud Computing in the Irish Construction Industry', *Proceeding CIB W78 2010, 27th International Conference,* 16-18 November, Cairo, Eygpt

Sebastian, R. (2010) 'Breaking through Business and Legal Barriers of Open Collaborative Process based on Building Information Modelling (BIM)', *Accepted paper for Proceedings of CIB World Congress,* 10-14 May 2010, Salford Quays, UK

Smith, D. (2007) 'An Introduction to Building Information Modeling (BIM)', *Journal of Building Information Modeling,* An official publication of the National BIM Standard (NBIMS) and the National Institute of Building Sciences (NIBS), 17, 12-14

Weinsten, B.D. (1993) 'What is an Expert?', *Theoretical Medicine and Bioethics,* **14** (1), 57-73

Wohl, A. (2008) '*Succeeding at SaaS: Computing in the Cloud*', Published by Wohl Associates

The Effects of the Global Financial Crisis on the Australian Building Construction Supply Chain

14

Ram Karthikeyan Thangaraj and Toong Khuan Chan (The University of Melbourne, Australia)

Abstract

This study involves a financial analysis of 43 publicly listed and large private companies in the building and construction supply chain from 2005 to 2010; straddling the period of the global financial crisis (GFC); and examines the impact of the GFC on the performance of these companies. The construction supply chain was divided into four sectors – material suppliers, construction companies, property developers and real estate investment trusts (REITs). The findings indicate that the impact was minimal for both material suppliers and construction companies, but especially severe for the more leveraged property developers and REITs. Building material suppliers and construction companies have benefitted substantially from the building economic stimulus packages provided by the Australian government to mitigate the effects of the GFC. Decreases in the valuation of assets have, to a large extent, reduced the profitability of property developers and REITs during the GFC but these companies have recovered quickly from these adverse conditions to return to a sound financial position by the end of the 2010 financial year. The results will inform investors, construction company managers and construction professionals in devising strategies for prudent financial management and for weathering future financial crises.

Keywords: Financial analysis, Business management, Construction, Global financial crisis, Australia

Introduction

The global financial crisis (GFC) is commonly believed to have begun in July 2007 when a loss of confidence by American investors in the value of sub-prime mortgages caused a liquidity crisis. By September 2008, the crisis had worsened with a sudden dramatic decline of stock prices around the globe. With a large number of borrowers defaulting on loans, banks were faced with a situation where the repossessed house and land was worth less on the market than the bank had loaned out originally. When Lehman Brothers collapsed in September 2008, governments around the world struggled to rescue large financial institutions as the fallout from the housing and stock market collapse worsened.

The Australian equity market was no different to many other countries in facing the GFC. The S&P/ASX200 index fell from a peak of 6700 in late 2007 to 3400 in November 2008. It was initially thought that Australia would fare much better as the local banks' exposure to collateralised debt obligations (CDOs) was relatively small in comparison with other countries. The housing market was also in a strong position with non-conforming loans in Australia accounting for only about 1% compared to 13% in the U.S. (Debelle 2008).

Despite the smaller exposure to toxic debts and a stronger housing market, the Australian gross domestic product (GDP) declined in the December quarter of 2008. Building construction commencements had reduced by 10% in the September 2008 quarter in response to the tightening of credit. Total building output for three subsequent quarters reduced as a consequence of the diminishing building commencements. Figure 1 below illustrates the change in GDP, building commencements and building output for each quarter between 2005 and 2010. The Australian property market was no different from any other industry facing the GFC. From the December 2007 quarter to the December 2008 quarter, owner occupied housing commitments fell by 22% (ABS 2010). This can be attributed to an increase in the unemployment rate during that period, and reduced confidence in the market. A market report states that office property developers in Australia experienced a 20.6% decline in their revenue for the financial year 2007-08 (IBISWorld 2011a). The retail market was also affected, resulting from reduced consumer spending.

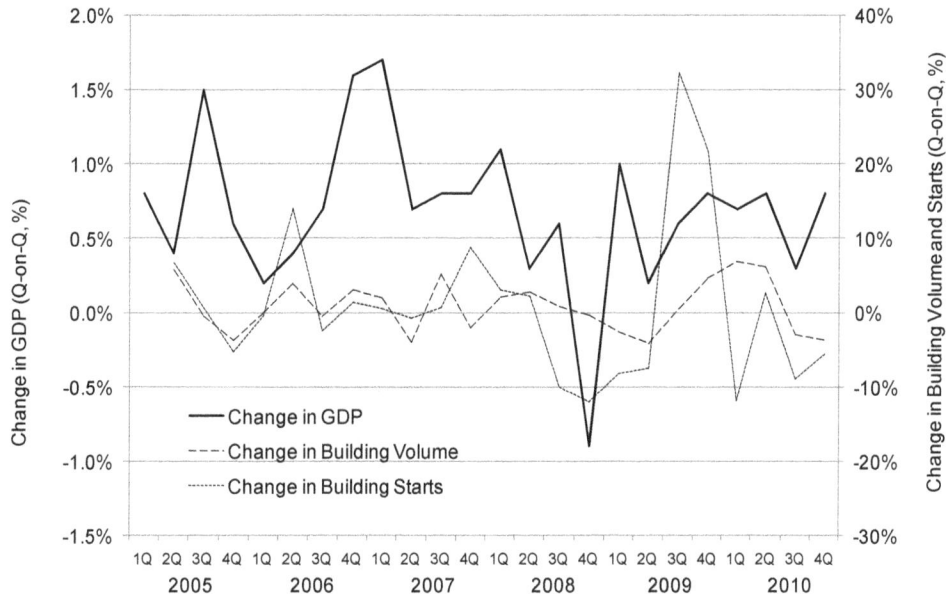

Figure 1 Changes in GDP, building starts and building volume (ABS 2010; ABS 2011)

Australia's Response to the GFC

The first significant macroeconomic policy response to the global financial crisis came from the Reserve Bank of Australia (RBA). On October 7, 2008, the RBA Board cut interest rates by 100 basis points to 6%. On October 12, the Australian Government announced it would guarantee all Australian bank deposits. Two days later, the Government announced an AUD10.4 billion stimulus package that included AUD1.5 billion to support housing construction. The housing aspect of the stimulus package – a time-limited grant to first home buyers – took effect immediately. To prepare for the possibility that the global financial crisis might be deeper and longer lasting than expected, the Government brought forward the commencement of large-scale infrastructure projects with a first tranche worth AUD4.7 billion announced in early December. On February 3, 2009 the Australian Government announced a second AUD42 billion stimulus package titled the Nation Building and Jobs Plan. As the global recession was expected to be deep and long, infrastructure spending played the central role with 70% of the second stimulus package to be spent on schools (AUD14.7 billion), social and defence housing (AUD6.6 billion), energy efficiency measures (AUD3.9 billion), and AUD890 million on road, rail and small-scale community infrastructure projects. More recent steps were contained in the 2009-10 Budget delivered in May 2009 where the Government introduced the third phase in its infrastructure program where an additional AUD22 billion was announced. The large increases in building starts in the third and fourth quarters of 2009 resulted from the initiation of stimulus packages earlier in the year. The Building the Education Revolution (BER), the single largest element of the Australian Government's Nation Building - Economic Stimulus Plan accounted for 23,600 projects.

These developments provided the motivation to develop a broader understanding of the effects of this recent financial crisis on the construction supply chain. The analysis focused on characterizing the impact of the GFC on the financial performance of companies within four sectors of the industry: building material suppliers, building contractors, property developers and Australian Real Estate Investment Trusts (A-REITs). The key points examined are the severity of the impact by means of ratio analysis, changes in cost structure and financial distress. The predicament of firms that were most affected was discussed in order to explore possible common characteristics of distress or failure. The findings will inform investors, managers and construction professionals in devising strategies for prudent financial management and for weathering future financial crises.

Background

Investigations into the financial performance of construction companies were initially reported in the 1970's. Sales turnover was identified to have a great impact on the profitability of the enterprise (Fadel 1977). A detailed industry level analysis of UK construction industry identified the similarities and differences in profits across general contractors and home builders (Akintoye and Skitmore 1991). Pre-tax profit margins among eighty general contractors hovered around 3.23% over the analysed period. Though uncorrelated with firm size, home builders' profit margin was observed to be four times more than general contractors. Low (1997) examined the performance of property companies in the Singapore stock exchange over a period of twenty one years and concluded that their performance was not better than the stock market and was highly correlated with the property market itself. Other researchers yielded the same findings in different countries (Abdul-Rasheed and Tajudeen 2006). On the contrary, Balatbat et al. (2011) reported that the performance of thirty Australian Securities Exchange (ASX) listed construction firms operating in civil infrastructure, residential and non-residential sectors over a 10 year period performed better than the All Ordinaries Index and a pool of blue chip companies. This was due to construction companies' higher efficiency in the utilisation of assets for revenue generation. A study (Cheah et al. 2004) on global construction firms indicates that business strategy has to account for efficiency in financial, operational, technical and human factors, and that corporate failure was largely due to failures in or two critical factors.

It is generally accepted practice to assess company performance using financial ratios – the analysis of these ratios over a period of time may provide substantial and reliable information on a company's financial health. Early statistical models based on these financial ratios were developed to analyse company failures. Beaver (1966) identified differences based on thirty financial ratios among groups of failed and non-failed firms as early as five years before the firms failed. The differences became more significant as firms approached failure. Altman (1968) combined ratio analysis and multiple discriminant analysis to formulate a statistical tool for prediction of company failure. This methodology analysed a combination of ratios, thereby eliminating possibilities of uncertainty in relying on single ratios. From a sample of sixty six manufacturing companies, of which half of them had filed for bankruptcy from 1946-1965, Altman derived a set of ratios covering liquidity, profitability, leverage, solvency and activity ratios. Many variations of this model have been created over time for specific industry segments in various regions: construction in the UK (Mason and Harris 1979; Abidali and Harris 1995), and for construction in China (Ng et al. 2011). Macroeconomic factors such as low construction activity, high interest rates, rise in inflation, and reduction in consumer spending have been identified by researchers to be significant factors that can drive companies towards failure (Kangari 1988; Arditi et al. 2000).

Both Beaver (1966) and Altman (1968) are of the view that financial ratios, based solely on financial statements, are an indicator of a company's past performance, and future performance is beyond prediction. Mason and Harris (1979) went on to declare that their model is able to identify companies that are 'at risk' of failure but cannot forecast whether a company will actually fail. It is worth pointing out at this stage that these tools only give an indication, and at any point changes in a company's strategy can transform a company from poor performance to market dominance. Langford et al. (1993) recognised that these analytical tools, along with the company's financial performance have to be used in combination for greater understanding of corporate performance. Similar views have been echoed by Chan et al. (2005) when analysing the impact of the Asian financial crisis on construction companies in Hong Kong.

Recent studies (Abdul-Rasheed and Tajudeen 2006, Chen 2009, Balatbat et al. 2010, Balatbat et al. 2011) have excluded building material suppliers and real estate investment trusts (REITs) in the review of financial performance, ignoring their importance within the building construction industry. An industry report estimates about 6,100 establishments to be operating in the materials supply sector in 2010-11 generating an estimated AUD20.5 billion in annual revenues (IBISWorld 2011b). Understanding the performance of the building material suppliers and REITs, in comparison with the builders and developers, is vital to the construction industry's resilience to future market downturns.

REITs came into existence in the US in 1960 (NAREIT 2011) to enable retail investors to invest in large scale real estate assets. Australian Real Estate Investment Trust (A-REIT) was established a decade later in 1971. The Australian Securities Exchange defines A-REITs as professionally managed and diversified portfolio of commercial real estate. Investors gain exposure to both the value of the real estate the trust owns, and the regular rental income generated from the properties.

Data and Methodology

In this study, the supply chain for building construction was limited to 4 sectors: building material suppliers, building contractors, property developers and A-REITs. A list of companies in each sector was determined by examining companies listed in the Australian Securities Exchange under Global Industry Classification Standards (GICS) capital goods and real estate industry groups. The next selection criteria for these companies were as follows to ensure that these companies were representative of the local economy; (i) listed in the ASX prior to 2006 financial year, (ii) more than 50% of their revenues were generated from Australian operations, and (iii) more than 50% of their revenue were generated from the sectors they represent, respectively. This filter resulted in a total of only 36 companies with only one operating exclusively as a building contractor. This is because most public listed building construction companies in Australia were concurrently operating in both construction and development. Hence in order to obtain a representative sample of the building contractors, the search was enlarged to include large (revenues exceeding AUD25 million) private construction companies that were required to file annual report to the Australian Security Investment Commission (ASIC). This search identified 51 companies of which only 16 had filed Form 388 for the years 2006 to 2010. Among the 16, only 7 companies were predominantly engaged in building construction. The other firms that were involved in either development or infrastructure works along with construction were not taken into consideration. The lack of local investment opportunities has seen Australian REITs seeking international property investments in 2000s. While this strategy brings about diversification gains, international property introduces additional currency, political and economic risks. Many of the A-REITs with large total assets were excluded from this analysis as these comprised significant international properties. A full list of the companies analysed is listed in Table 1.

The financial performance of all selected companies was calculated over the 2006-10 period, both years inclusive. Despite its many drawbacks, the most precise information on these companies can only be determined by relying on audited financial statements (Langford et al. 1993). In order to assess the impact of the GFC on the performance of these companies, the values for every consecutive year was compared in a trend analysis. The cost structure for the four sectors has been examined with cost of goods sold (COGS), expenses, depreciation, amortization, interest and profit displayed as a percentage of revenue. Items such as rent, utilities, wages and purchases were not always discernable or available and were considered as part of expenses. A distress analysis was conducted using the Altman's Z-index (1968). This was considered to be the most appropriate model to compare the risk of failure across the four different sectors.

Material Supplier	Building Construction
Adelaide Brighton Limited (ASX:ABC) Alesco Corporation Limited (ASX:ALS) Boral Limited (ASX:BLD) Bluescope Steel Limited (ASX:BSL) Brickworks Limited (ASX:BKW) CSR Limited (ASX:CSR) Gunns Limited (ASX:GNS) GWA Group Limited (ASX:GWA) Reece Australia Limited (ASX:REH)	Badge Constructions (SA) Pty Ltd Hooker Cockram Corporation Pty Ltd J. Hutchinson Pty Ltd Masterton Corporation Holding Company Pty Ltd Pellicano Builders Pty Ltd Reed Constructions Australia Pty Ltd St.Hilliers Construction Pty Ltd Tamawood Limited (ASX:TWD)
Property Developer	**AREIT**
AHC Limited (ASX:AHC) Australand Property Group (ASX:ALZ) AVJennings Limited (ASX:AVJ) Becton Property Group (ASX:BEC) Cedar Woods Properties Ltd (ASX:CWP) CIC Australia Limited (ASX:CNB) Devine Limited (ASX:DVN) Finbar Group Limited (ASX:FRI) FKP Property Group (ASX:FKP) Geo Property Group (ASX:GPM) International Equities Corp Ltd (ASX:IEQ) Lend Lease Group (ASX:LLC) Metroland Australia Limited (ASX:MTD) Mirvac Group (ASX:MGR) Payce Consolidated Limited (ASX:PAY) Peet Limited (ASX:PPC) Stockland (ASX:SGP) Sunland Group Limited (ASX:SDG)	Abacus Property Group (ASX:ABP) APN Property Group Limited (ASX:APD) Aspen Group (ASX:APZ) BWP Trust (ASX:BWP) CFS Retail Property Trust (ASX:CFX) Charter Hall Group (ASX:CHC) Commonwealth Property (ASX:CPA) Cromwell Property Group (ASX:CMW) Goodman Group (ASX:GMG) GPT Group (ASX:GPT) Trafalgar Corporate Group (ASX:TGP)

Table 1 List of Companies studied

Performance Analysis of Companies

The best gauge for measure of a recession, next to GDP is the stock market. The market index is a reliable tool to plot the timeline of recession right from start to recovery. The primary investable benchmark in Australia, the S&P/ASX200, fell from its peak of 6748 in October 2007 to 3145 in March 2009, losing 46% of its valuation. Since then the markets have recovered and remained fairly buoyant at 4500 until the end of 2010. With the market as a reliable pointer, the period of analysis has been set from financial year 2006 to 2010 to cater for the boom, fall and recovery due to the GFC in Australia.

In terms of revenue, property developers and A-REITs exhibited significant decline in revenues in 2008 and 2009, and in the case of developers, continuing into 2010. The building material sector exhibited a marginal contraction of 2.8% in 2009 followed by a further reduction of nearly 7% in 2010. On the contrary, revenues for building contractors continued to increase, albeit by only 1.6% in 2009 and 0.8% in 2010 despite the slump in building starts in late 2008. The nature of the building construction business is that builders are not immediately affected by a downturn due to continuing construction projects awarded a couple of years earlier. The launch of a number of stimulus packages; bringing forward spending on large-scale infrastructure and additional spending on new school buildings in late 2008 and early 2009 by the Australian government to mitigate the effects of the GFC on the local construction market have maintained building starts at a level of AUD81.5 billion and AUD75.6 billion in 2008 and 2009, respectively (ABS 2010; ABS 2011).

Table 2 indicates that the total net assets of all the companies analysed have increased in the period examined. In fact, net assets for building contractors have more than doubled compared to 2006 with other sectors averaging an increase of 50%. This is not withstanding the small decreases in net assets for property developers and A-REITs that had to revalue some of their assets in 2009 and 2010. Net asset in this table is usually equal to the shareholders' equity in the company balance sheet or total assets minus total liabilities.

Year	2006	2007	2008	2009	2010
Sales Revenue					
Building Material Suppliers	19,926	20,961	24,454	23,775	22,136
(Change, % year-on-year)		5.2%	16.7%	-2.8%	-6.95
Building Contractors	1,506	2,068	2,455	2,495	2,514
(Change, % year-on-year)		37.3%	18.7%	1.6%	0.8%
Property Developers	7,205	9,323	7,827	6,903	6,810
(Change, % year-on-year)		29.4%	-16.1%	-11.8%	-1.4%
A-REITs	1,636	2,762	2,294	1,331	1,493
(Change, % year-on-year)		68.8%	-16.9%	-42.0%	12.2%
Net Assets					
Building Material Suppliers	10,404	11,936	12,725	15,105	15,731
(Change, % year-on-year)		14.7%	6.6%	18.7%	4.1%
Building Contractors	152	248	301	329	357
(Change, % year-on-year)		63.1%	21.2%	9.5%	8.4%
Property Developers	13,137	17,676	18,760	18,693	19,705
(Change, % year-on-year)		34.6%	6.1%	-0.4%	5.4%
A-REITs	6,187	8,102	10,134	9,521	9,132
(Change, % year-on-year)		30.9%	25.1%	-6.0%	-4.1%

Table 2 Sales revenue and Net assets (AUD million) for years 2006 to 2010

Ratios measuring profitability, liquidity, activity, leverage and solvency are generally accepted as the most significant indicators of corporate performance. Table 3 lists at least one financial ratio in each of the categories above to evaluate the performance of the companies in the building construction supply chain. These ratios were weighted based on the annual revenues of the respective companies. In terms of profitability, the net profit margin (net profit divided by sales revenue) for material suppliers reduced progressively from more than 7% in 2006 to 0.8% in 2010. Net profit margin for building contractors was low at between 3-4% in 2006 and 2007, and reduced to a minimum of 1.6% during the depths of the financial crisis. It recovered slightly to 1.9% in 2010. The net profit margin for property developers was more than 30% before the onset of the GFC, but converted into a 59% loss in 2009. Similarly, A-REITs reported a net profit margin of 70% and 53% pre-GFC and a massive loss of 106% in 2009. Both the property developer and AREIT sectors returned to profitability and recovered to approximately one half their pre-GFC margins in 2010.

Return on average equity (ROE) measures the rate of return on the shareholders' equity and reflects the company's efficiency at generating profits from every dollar of equity. Defined as net profits divided by average equity, the building material supplier sector exhibited returns between 13% and 17% during the period examined but dropped to a low of 0.2% during 2009 when net profits were depressed by the effects of the GFC. Pre-GFC, the building contractors were achieving returns in excess of 40% but fell to 23.1% in 2008 and eventually to a low of 8.3% in 2010. This corresponded with the observed drop in net profit margin from more than 30% pre-GFC to less than 20% post-GFC. The property developers achieved an ROE of approximately 20% pre-GFC but suffered losses in 2009 to report an ROE of -24% during 2009 and recovering to a small 0.4% in 2010. The ROE for the A-REIT sector fell from 21% in 2006 to -13.9% at the pits of the GFC in 2009 but recovered when the sector returned to profitability in 2010.

The current ratio, defined as current assets divided by current liabilities, provides a measure of liquidity or the company's ability to pay back its short-term liabilities with cash, inventory and receivables. As can be seen from Table 3, the current ratios for both the building material suppliers and building contractors were between 1.5 and 1.9, indicating that these companies were adequately liquid and able to pay its obligations. The current ratios remained above 1.0 during the period of the financial crisis indicating that these companies were managing their cash flows prudently. A number of companies invested in additional property, plant and equipment, increasing debts and consequently reducing the current ratio slightly in 2008. Once the effects of the GFC were apparent, these companies sought ways to reduce their current liabilities with a slight improvement in current ratio in 2009. Property developers exhibited an increase in current ratios in 2007-08 as completed and unsold properties were accounted for under current inventories leading to a higher current asset value. Impairments to the value of properties in inventories and the increases in interest bearing liabilities resulted in higher current liabilities. These changes led to current ratios fluctuating between 1.4 and 2.3 during this period. The current ratio for AREITs ranged from 0.66 to 1.33 during this period, but was not considered further as this sector invested primarily in properties - a non-current asset.

Financial Year	2006	2007	2008	2009	2010
Building Material Suppliers					
Net profit margin	0.074	0.084	0.062	0.015	0.014
Return on Average Equity	0.149	0.173	0.134	0.002	0.012
Current Ratio	1.437	1.544	1.257	1.742	1.780
Working Capital Turnover	11.760	9.430	54.230	8.224	6.770
Quick Ratio	0.867	0.933	0.722	0.944	0.988
Debt Ratio	0.540	0.491	0.524	0.432	0.415
Times Interest Earned Ratio	93.172	12.231	10.252	3.604	8.059
Building Contractors					
Net profit margin	0.034	0.040	0.028	0.016	0.019
Return on Average Equity	0.477	0.439	0.231	0.122	0.083
Current Ratio	1.479	1.504	1.472	1.636	1.466
Working Capital Turnover	6.712	4.650	15.489	11.381	50.254
Quick Ratio	1.104	1.018	1.034	1.181	1.074
Debt Ratio	0.683	0.653	0.654	0.599	0.651
Times Interest Earned Ratio	1525	6295	10194	816	48
Property Developers					
Net profit margin	0.324	0.377	0.153	-0.587	0.144
Return on Average Equity	0.203	0.189	0.069	-0.237	0.004
Current Ratio	1.393	1.725	1.656	2.306	1.460
Working Capital Turnover	-2.214	0.911	-1.684	2.270	3.459
Quick Ratio	0.454	0.778	0.602	1.367	0.726
Debt Ratio	0.534	0.479	0.502	0.463	0.428
Times Interest Earned Ratio	20.462	15.609	5.621	-15.014	8.088
REITs					
Net profit margin	0.707	0.776	0.581	-1.057	0.371
Return on Average Equity	0.210	0.222	0.134	-0.139	0.048
Current Ratio	0.663	1.049	0.750	1.331	0.736
Working Capital Turnover	-1.227	-0.839	-0.083	-0.386	0.787
Quick Ratio	0.643	0.897	0.702	1.210	0.646
Debt Ratio	0.357	0.336	0.358	0.334	0.330
Times Interest Earned Ratio	10.663	13.324	8.294	-9.515	2.749

Table 3 Financial ratios (weighted on revenue)

The quick ratio, a more severe measure of liquidity, defines the company's ability to meet its short-term obligations with its most liquid assets. Inventory was excluded from the current assets because of the delays involved in turning inventory into cash. As expected, the quick ratios for material suppliers were less than the current ratio, but were observed to be slightly below 1.0 indicating that current liabilities were marginally higher than current assets. The increase in short term debt in 2008 resulted in a large drop in quick ratio to 0.722. Once these liabilities were restructured in response to the GFC, the quick ratio improved to previous levels. Recognising that building contractors operate on very tight but prudent cash flows, the quick ratios for this sector were all marginally above 1.0 during the entire period examined. The property developers and AREITs exhibited notably low quick ratios as properties were considered to be non-current assets. Again, some of these companies had to restructure their liabilities in 2009 leading to improved quick ratios for the year.

The working capital turnover (WCT) ratio is an activity ratio which represents the number of times the working capital is turned over in the course of a year. A higher WCT ratio indicates the efficiency with which the working capital is being used to generate revenue. The building material suppliers exhibited a WCT ratio between 6.8 and 11.8 for the period examined except for 2008 when the ratio jumped to a high of 54. This was due to large current liabilities incurred by a number of companies leading to a reduced working capital. The large WCT ratio for building contractors between 2008 and 2010 was also due to reductions in working capital as the revenue was previously reported to be relatively constant over this period. The WCT ratio for both the property developers and AREITs did not provide much information as the values fluctuated between -2.2 and +3.5, most likely attributed to large changes in the levels of current assets and liabilities.

The use of debt to finance the company's assets was evaluated by examining the debt ratio which is defined as the ratio of total debt to total assets. The debt ratios remained fairly constant over the entire period examined for all four sectors although the building contractors showed the highest debt ratio at 65% whereas the AREITs had the lowest ratio of approximately 33%. Building contractors exhibited comparatively higher debt ratios due to high levels of short term trade debts payable to sub-contractors. When the debt ratio was examined together with the Times Interest Earned (TIE) ratio, it clearly indicated that although the building contractors were most highly leveraged, the debt amounts were small relative to the net profits earned by these companies. Building contractors were obviously able to meet its debt obligations many times over. It can be inferred from an examination of the combination of debt and the TIE ratios that most of the liabilities of the AREITs business structure are long term debts whereas building contractors incur short term debts. The TIE ratio for building material suppliers reduced drastically from 93 in 2006 to 3.6 in 2009 due to a drop in net profits. The effect of the GFC was evident in two sectors; the property developers and AREITs, where a majority of these companies suffered losses in 2009 leading to negative TIE ratios.

Industry Cost Structure

The industry cost structure was examined by comparing input costs to revenue for each sector. The cost of input materials (represented by cost of goods sold, COGS) and operating expenses are a significant component of the material supplier cost structure. From Figure 2a it can be observed that COGS and expenses have remained fairly consistent at 56-58% and 29-30%, respectively, leading up to the GFC. Other costs such as depreciation, amortization and interest expense accounts for very small percentages of the total revenue. The most significant effect of the GFC is the drop in profit from 8% in 2007 to 1% in 2010.

For the building construction sector, COGS represents close to 90% of total revenue with profits accounting for 3-4% pre-GFC as illustrated in Figure 2b. It can be seen that all the inputs have remained relatively stable with an evident squeeze on profits. Interest expense for the building construction sector is lowest amongst all four sectors at less than 0.1%

indicating that companies in this sector have insignificant bank loans. The cost structure for property developers include 70% for cost of goods and another 12-15% for operating expenses. Interest payments amount to approximately 3% of total revenues. In terms of profits, property development sector have reported large swings in figures in excess of 12% profit pre-GFC to losses amounting to approximately 20% of revenue in 2009. The increase in expenses was attributed to a write-down in the value of property assets held by these companies during the construction stage. It may be worthwhile to note that this sector recovered in 2010 reporting a profit of 7%.

Property investment companies and REITs have been relatively quick to react to the GFC compared to the other sectors in the supply chain. This is clearly due to a drop in value of the financial assets held by the investment companies and trusts. REITs operate in a similar manner as investment funds and do not report COGS. Interest payments amount to approximately 10% as this sector is heavily reliant on debt to finance the purchase of real estate assets. Figure 2d clearly indicates that REITs were amongst the most severely affected with a reported cost structure nearly three times that of the total revenue for 2009. This sector exhibited the highest profit margins with pre-GFC profit levels at 66% of total revenue. A decline was seen in the 2008 financial year, when the effects of the GFC started to affect the bottom line with a loss of 27%, and continuing the decline with a loss amounting to nearly 200% of total revenue in 2009. The effect of the GFC was two-fold: a loss of confidence in the property market meant that real estate valuations plunged requiring these investment companies and trusts to write-down assets values; and a drop in occupancy leading to a sharp decline in revenue.

Distress Analysis
Altman (1968) used the multiple discriminant analysis model to combine several financial ratios into a single index, called the 'distress score' to discriminate between failed and no-failed groups. Altman's proposed Z-score, based on an analysis of 66 manufacturing companies of which 33 had failed in 1946, estimated the risk of bankruptcy within a period of two to three years. This analysis was performed to investigate the possibility of business failures in response to the GFC.

Table 4 indicates that building contractors have remained fairly stable and well above the 'at risk' score of 2.7 throughout the period of study. Material suppliers on the other hand exhibited excellent scores for 2006-08 but dipped slightly into the 'at risk' score for 2009-10 having suffered a drop in revenue and market value. Property developers and REITs were categorised as 'at risk' before the onset of the GFC but fell below the 'imminent failure' threshold of 1.8 in 2008. On examining the individual factors for the Z-score, it is clear that the fall from the safe level for property developers and REITs in 2008 has been influenced by a loss in profitability and productivity.

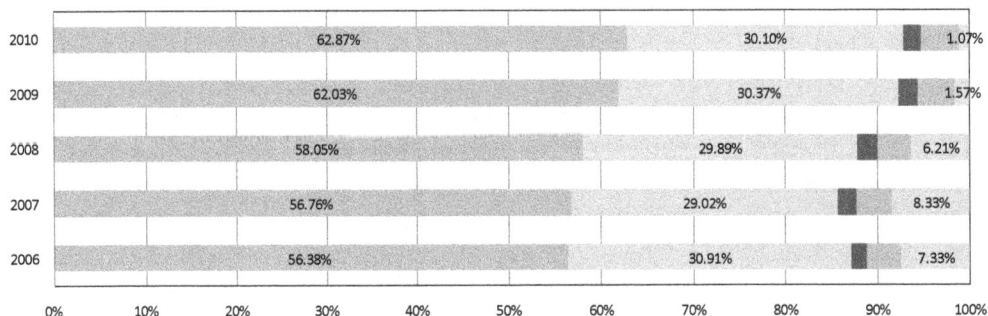

(a) Material suppliers

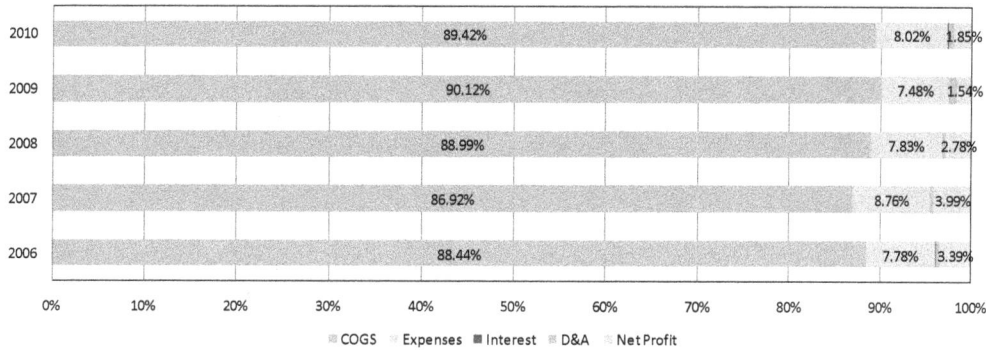

(b) Construction

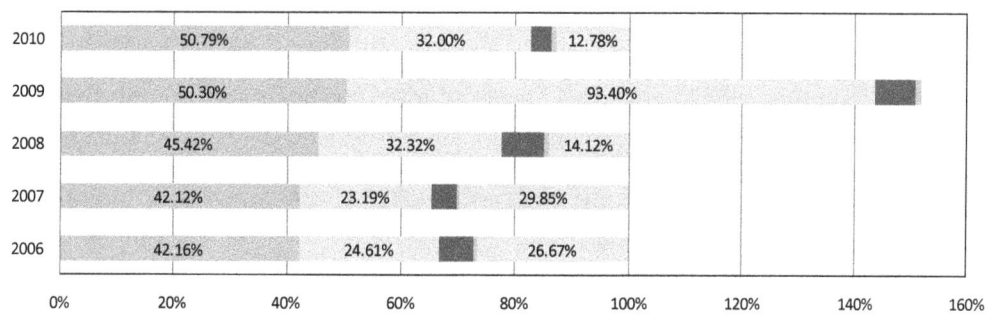

(c) Property developers

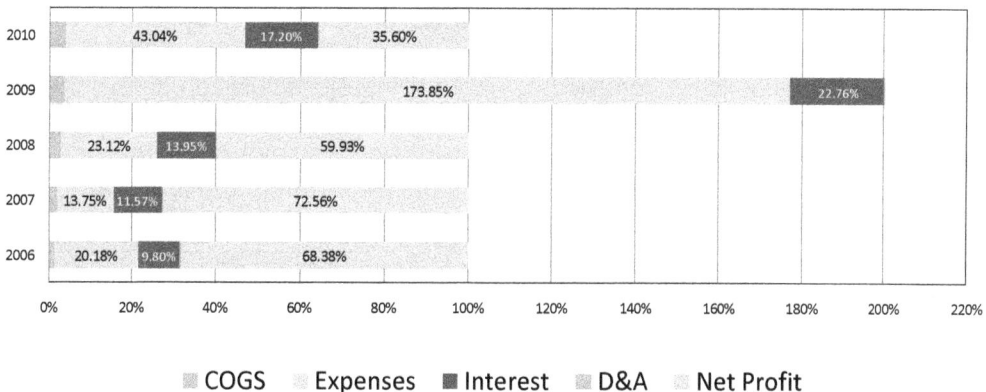

■ COGS   Expenses   ■ Interest   D&A   Net Profit

(d) REITs

Figure 2 Cost structure for material suppliers, construction, property developers and REIT sectors (weighted based on revenue)

Property developers and REITs with higher returns initially have seen greater erosion in their profitability to the extent of negative earnings in 2009 whereas building supplies and building contractors have remained profitable throughout the period. For both developers and REITs, interest expenses, depreciation and amortization have remained fairly stable throughout the period. The reduction in profitability was significantly influenced by decreases in valuation of their assets and financial derivatives on hold, and impairment charges.

The distress analysis indicates that the building materials and contractors in Australia were financially sound and were able to accommodate the impact of the GFC. A similar analysis of contractors in Hong Kong (Chan et al. 2005) reported Z-scores falling from 2.33 in 1997/98 to 1.41 in 2001/02 due to severe competition and reduced demand for property.

Financial Year	2006	2007	2008	2009	2010
Building Material Suppliers					
Z-score	2.79	3.40	2.83	2.22	2.50
Building Contractors					
Z-score	4.97	4.73	4.35	4.69	4.00
Property Developers					
Z-score	1.72	2.05	1.14	0.26	1.11
REITs					
Z-score	2.43	2.88	1.40	0.77	1.29

Table 4 Distress analysis (weighted on revenue)

Discussion of Results

The results indicate that each of the four sectors of the building supply chain was affected to different extents by the GFC. The building material suppliers were the least affected exhibiting a 3% year-on-year drop in revenue in 2009 and another 7% drop in 2010. These companies remained profitable albeit at a much reduced profit margin. Net assets of these companies continued to grow over the entire period of study. A number of these companies were in the business of manufacturing products for other sectors of the economy and were marginally affected by the decline in construction demand in 2008. The activity ratio for these companies remained relatively constant over the period. These companies were moderately leveraged with debt, but earned sufficient revenues to cover the interest payments many times over.

The building contractors sector exhibited increases in revenue despite the drop in total building starts in 2008 and the consequential drop in building volumes in 2009. This reflects the mode of operation of building contractors where the project duration delays the effects of a downturn to a couple of years after the onset of a fall in building starts. The net assets of the building contractors, although lowest of all four sectors, have more than doubled from 2006 to 2010, indicating that substantial additional investments were made into these companies. Profit margins which were initially low at 3% to 4% were further depressed by the competition for jobs during the lead up to the GFC. The profit margins shown in Table 3 for the building contractor sector were the lowest of all sectors examined. Although the profit margin was low, this sector remained profitable and financially liquid throughout the GFC period. The observed profit margin was not much different from the margin of 3.23% reported by Akintoye and Skitmore (1991) for an analysis of 80 general contractors in the UK. It also compares reasonable better than the 1% to 2% reported by Chan et al. (2005) for construction companies in Hong Kong immediately after the 1997 Asian Financial Crisis. The ROE ratios in excess of 40% confirms the perception that construction activities require relatively lower amounts of equity and that operations is normally funded through judicious utilisation of working capital. Construction companies, though exhibiting debt ratios approximately 60%, are able to repay the interests on their debts from their profits. In general, the financial ratios indicate that they have a well-founded financial position, and are not directly prone to demand fluctuations are generally perceived. A recent study by Balabat (2011) of thirty ASX-listed construction firms operating in the civil infrastructure, residential and non-residential sectors reported that these firms exhibited a 25% higher growth in share value compared to blue chip companies and 48% higher compared to the ASX-All Ordinaries Index.

Apparently, both the building material supplier and building contractor sectors benefitted from the stimulus packages once the numerous building projects commenced in earnest in mid- to late-2009. These sectors remained profitable and solvent, exhibited insignificant changes to leverage levels; the only indication of concern was a drop in net profits and interest coverage.

Property developers generally rely on internally generated funds and borrowings to finance their investment opportunities resulting in the largest net assets reported for the four sectors as shown in Table 2. Given that these developers in Australia were making healthy profits on the back of the property boom before the onset of the GFC, these companies tend to utilise more long-term debt to finance their business operations. Although the year-on-year falls in revenue between 2007 and 2009 were only 16% and 12%, respectively, a drop in the valuation of properties held or under-construction resulted in serious losses in 2009. Once the market recovered partially in 2010 and the revaluation of properties was conducted, these companies reported net profits and reverted to a healthy balance sheet.

The REITs in Australia engage in the acquisition and ownership of property and primarily derive their income from rental or leasing whereas property developers are mainly involved in the development and management of real estate properties. Given that rental revenues are usually relatively stable, REITs tend to be perceived as a low risk investment vehicle. This study observed net profit margins in excess of 70% during 2006 and 2007, before the onset of the GFC, to a substantial loss of 106% of revenue reported in the 2009 financial year for the REITs sector. However, a revaluation of its property portfolio post-GFC resulted in net profit margins recovering to 37% in 2010. Although property portfolios of the developers and REITs were similarly assessed each financial year, the impact of any upward and downward swings was greater for the REITs sector than the property development sector. Among the sub sectors being analysed, REITs were in a much better situation to cover for their current liabilities with the lowest debt ratios of 0.33 to 0.36. As the REITs' business risk is relatively lower than the other three sectors, these companies utilise more long-term debts to finance their investments in real estate. It may be worthwhile to note that the sample of REITs shown here have reduced the debt ratio from 0.36 in 2006 to 0.33 in 2010. Excessive levels of gearing have magnified the steep decline in asset values for a number of REITs during the GFC.

Conclusions

The financial analysis of the companies operating across the building construction supply chain in Australia indicates that all four sectors examined were susceptible to the global financial crisis. From the ratio and cost structure analyses, it was evident that building material suppliers were the least affected as these companies have remained profitable and solvent throughout the period examined. These companies have benefited substantially from the building economic stimulus package provided by the Australian government. The impact on building contractors was also limited, as they have also remained profitable albeit at significantly reduced levels. However, there is a very high likelihood that the findings may be different if the study period was extended beyond 2010 as the projects let under the stimulus package are completed.

Developers and REITs were the most affected sectors in this financial crisis where significant erosion in profitability was observed. The most severe drop in profitability was reported in 2009 when developers and REITs exhibited losses amounting to 25% and 105% of revenue, respectively. Decreases in the valuation of their assets and financial derivatives on hold, and impairment charges, have played a major role in reducing their profitability. The post-GFC recovery to profitability has been triggered by an increase in their asset valuation, more specifically, owned properties and those for sale in inventories. Since 2008, the Z-scores for developers and REITs have fallen below the 'distress' level, but it was encouraging to see an increase from 2009 and 2010 indicating that the worse was over. The GFC eroded the peak market capitalisation of Australian REITs from AUD147 billion in 2007 to its present value of AUD74 billion at the end of the 2010 financial year (ASX 2007; ASX 2011). The steep decline of asset values during the GFC was magnified by the high gearing employed by the A-REIT sector. The level of debt has reduced slightly to 0.33 in 2010 indicating a more conservative and managed approach to gearing in this sector after the GFC.

Post Script

After this research project was completed, two companies in the list shown in Table 1 – Reed Constructions and St Hilliers Construction, were reported to be in financial distress.

Reed Constructions, based in NSW, was reported to be in financial distress, with up to $80 million of bills outstanding, despite being allocated $383.3 million under the Building the Education Revolution program (Moran, 2012a). On 15 June 2012, Reed Constructions Australia Pty Limited was placed in Voluntary Administration (Ferrier Hodgson, 2012) where debts were reportedly at $182.1 million after suffering losses on several other government contracts. On 9 July 2012, Reed Constructions was placed in the hands of a liquidator following a successful winding up application filed in the NSW Supreme Court by a creditor, SCE Group (Schlesinger, 2012). Data for Reed Constructions indicate that revenues increased at a compounded annual growth rate of 25% from $171 million in 2006 to $466 million in 2011 with profit before tax at $18 million and $24 million, respectively for these two financial years. At the end of the 2011 financial year, net assets were at $99 million, quick ratio at 1.12 and the debt ratio at 0.54 indicating that the company was reasonably healthy at that point in time. The 2011 financial statements indicate a $47.1 million net cash outflow from its construction activities.

St Hilliers Construction Pty Limited was placed in Voluntary Administration on 16 May 2012 after a dispute over the funding of the $350 million Ararat Prison project that it was constructing under a public-private partnership approach (Zappone, 2012). It was reported that St Hillers Construction owed its creditors and employees a total of $121 million (Moran, 2012b). The financial statements indicate that although the revenues increases from $347 million in 2006 to $515 million in 2008 and declined slightly to $476 and $492 million in 2009 and 2010, respectively, the company returned profits between $5 million and $12 million between 2006 to 2009. The company exhibited early signs on distress when it incurred a loss of nearly $18 million, before tax, for the 2010 financial year mainly due to higher construction costs (or conversely, lower margins). Net assets remained positive as $27 million, but its quick ratio deteriorated to 0.54 in 2010 compared to 1.00 the previous year. It suffered a net cash outflow of $16.0 million in 2010 and a further $13.3 million in the 2011 financial year.

These two recent cases of financial distress is the subject of an ongoing study to further characterise the performance of the building and construction sector in these uncertain financial conditions: the results will follow in a future publication. Preliminary data seems to suggest that an adequate profit margin is required to provide a buffer against unforeseen project risks. The continued effects of the GFC will become more evident once financial statements and other pertinent corporate information are made available in the public domain.

References

Abidali, A.F. and Harris, F. (1995) 'A Methodology for Predicting Company Failure in the Construction Industry', *Construction Management and Economics*, **13**, 189-196

Abdul-Rasheed, A. and Tajudeen, A.B. (2006) 'Performance analysis of listed construction and real estate companies in Nigeria', *Journal of Real Estate Portfolio Management*, **12** (2),177-185

Australian Bureau of Statistics (2010) '8752.0 Building Activity, December Quarter 2009', April 2010

Australian Bureau of Statistics (2011) '5206.0 Australian National Accounts: National Income, Expenditure and Product, June Quarter 2011', September 2011

Akintoye, A. and Skitmore, M. (1991) 'Profitability of UK Construction Companies', *Construction Management and Economics*, **9** (4), 311-325

Altman, E.I. (1968) 'Financial ratios, discriminant analysis and the prediction of corporate bankruptcy', *The Journal of Finance*, **23** (4), 589-609

Arditi, D., Koksal, A., and Kale, S. (2000) 'Business Failures in the Construction Industry', *Engineering, Construction and Architectural Management*, **7** (2), 120-132

Australian Securities Exchange ASX. (2007) 'Listed Managed Investments (LMI) – Monthly Update – August 2007'

Australian Securities Exchange (2011) 'Listed Managed Investments (LMI) – Monthly Update – August 2011'

Balatbat, M.C., Lin, C-Y. and Carmichael, D.G. (2010) 'Comparative performance of publicly listed construction companies: Australian evidence', *Construction Management and Economics*, **28**, 919-932

Balatbat, M.C., Lin, C-Y. and Carmichael, D.G. (2011) 'Management Efficiency Performance of Construction Business: Australian Data', *Engineering, Construction and Architectural Management*, **18** (2), 140-158

Beaver, W.H. (1966) 'Financial Ratios as Predictors of Failure', *Journal of Accounting Research*, **4**, 71-111

Chan, J. K., Tam, C. M., and Cheung, R. (2005) 'Monitoring Financial Health of Contractors at the Aftermath of the Asian Economic Turmoil: A Case Study in Hong Kong', *Construction Management and Economics*, **23**, 451-458

Cheah, C.Y.J., Garvin, M.J. and Miller, J.B. (2004) 'Empirical Study of Strategic Performance of Global Construction Firms', *Journal of Construction Engineering and Management*, **130** (6), 808-817

Chen, H.L. (2009) 'Model for Predicting Financial Performance of Development and Construction Corporations', Journal of Construction Engineering and Management, **135** (11), November 2009

Debelle, G. (2008) 'A Comparison of the US and Australian Housing Markets. Reserve Bank of Australia Bulletin', June 2008

Fadel, H. (1977) 'The Predictive Power of Financial Ratios in the British Construction Industry', *Journal of Business Finance & Accounting*, **4** (3), 339-352

Ferrier Hodgson (2012) 'Reed Constructions Australia Placed in Administration', Media Release, Ferrier Hodgson, 15 June 2012, Retrieved August 28, 2012, from http://www.ferrierhodgson.com

IBISWorld Pty Ltd. (2011a) 'Office Property Operators in Australia', Melbourne

IBISWorld Pty Ltd. (2011b) 'Building Supplies Wholesaling in Australia', Melbourne

Kangari, R. (1988) 'Business Failure in the Construction Industry', *Journal of Construction Engineering and Management*, **114** (2), 172-190

Langford, D., Iyagba, R, and Komba, D. (1993) 'Prediction of Solvency in Construction Companies', *Construction Management and Economics*, **11** (5), 317-325

Low, K. H. (1997) 'The Historical Performance of Singapore Property Stocks', *Journal of Property Finance*, **8** (2), 111-125

Mason, R. and Harris, F. (1979) 'Predicting Company Failure in the Construction Industry', *Proceedings Institution of Civil Engineers*, **66**, 301-307

Moran, S. (2012a) 'Probe into Reed's handling of $90m for BER', *The Australian*, 6 March 2012, Retrieved August 28, 2012, from http://www.theaustralian.com.au/national-affairs/state-politics/probe-into-reeds-handling-of-90m-for-ber/story-e6frgczx-1226289882027

Moran, S. (2012b) 'Administrators Moore Stephens seeking buyer for St.Hilliers by August', The Australian, 14 June 2012, Retrieved August 28, 2012, from http://www.theaustralian.com.au/business/property/administrators-moore-stephens-seeking-buyer-for-st-hilliers-by-august/story-fn9656lz-1226394814547

National Association of Real Estate Investment Trusts (2011) *'REIT Industry Timeline: Celebrating 50 years of REITs and NAREIT'*, Retrieved September 1, 2011, from http://www.reit.com/timeline/timeline.php

Ng, S.T., Wong, J.M. and Zhang, J. (2011) 'Applying Z-Score Model to Distinguish Insolvent Construction Companies in China', *Habitat International*, **35**, 599-607

Schlesinger, L. (2012) 'Liquidator appointed for collapsed NSW builder Reed Construction', Property Observer, 11 July 2012, Retrieved August 28, 2012, from http://www.propertyobserver.com.au/industry-news/liquidator-appointed-for-collapsed-nsw-builder-reed-construction/2012071155520

Zappone, C. (2012) 'St Hilliers placed in voluntary administration after jail break', The Sydney Morning Herald, 16 May 2012, Retrieved August 28, 2012, from http://www.smh.com.au/business/st-hilliers-placed-in-voluntary-administration-after-jail-break-20120516-1yprv.html

Permissions

All chapters in this book were first published in AJCEB, by UTS ePress; hereby published with permission under the Creative Commons Attribution License or equivalent. Every chapter published in this book has been scrutinized by our experts. Their significance has been extensively debated. The topics covered herein carry significant findings which will fuel the growth of the discipline. They may even be implemented as practical applications or may be referred to as a beginning point for another development.

The contributors of this book come from diverse backgrounds, making this book a truly international effort. This book will bring forth new frontiers with its revolutionizing research information and detailed analysis of the nascent developments around the world.

We would like to thank all the contributing authors for lending their expertise to make the book truly unique. They have played a crucial role in the development of this book. Without their invaluable contributions this book wouldn't have been possible. They have made vital efforts to compile up to date information on the varied aspects of this subject to make this book a valuable addition to the collection of many professionals and students.

This book was conceptualized with the vision of imparting up-to-date information and advanced data in this field. To ensure the same, a matchless editorial board was set up. Every individual on the board went through rigorous rounds of assessment to prove their worth. After which they invested a large part of their time researching and compiling the most relevant data for our readers.

The editorial board has been involved in producing this book since its inception. They have spent rigorous hours researching and exploring the diverse topics which have resulted in the successful publishing of this book. They have passed on their knowledge of decades through this book. To expedite this challenging task, the publisher supported the team at every step. A small team of assistant editors was also appointed to further simplify the editing procedure and attain best results for the readers.

Apart from the editorial board, the designing team has also invested a significant amount of their time in understanding the subject and creating the most relevant covers. They scrutinized every image to scout for the most suitable representation of the subject and create an appropriate cover for the book.

The publishing team has been an ardent support to the editorial, designing and production team. Their endless efforts to recruit the best for this project, has resulted in the accomplishment of this book. They are a veteran in the field of academics and their pool of knowledge is as vast as their experience in printing. Their expertise and guidance has proved useful at every step. Their uncompromising quality standards have made this book an exceptional effort. Their encouragement from time to time has been an inspiration for everyone.

The publisher and the editorial board hope that this book will prove to be a valuable piece of knowledge for researchers, students, practitioners and scholars across the globe.

List of Contributors

Saad Sarhan
University of Plymouth, UK

Andrew Fox
University of Plymouth, UK

Brian Hubbard
Purdue University, USA

Qian Huang
Purdue University, USA

Patrick Caskey
Purdue University, USA

Yang Wang
Purdue University, USA

Ryan Stanley
Unitec Institute of Technology, New Zealand

Derek Thurnell
Unitec Institute of Technology, New Zealand

Thayaparan Gajendran
The University of Newcastle, Australia

Graham Brewer
The University of Newcastle, Australia

Malliga Marimuthu
Universiti Sains Malaysia, Malaysia

Lena Borg
Royal Institute of Technology, Sweden

Hans Lind
Royal Institute of Technology, Sweden

Perry Forsythe
School of the Built Environment, University of Technology Sydney, Australia

Abayomi Ibiyemi
University of Malaya, Malaysia

Yasmin Mohd Adnan
University of Malaya, Malaysia

Md Nasir Daud
University of Malaya, Malaysia

Martins Adenipekun
Lagos State Polytechnic, Nigeria

Ayokunle Olubunmi Olanipekun
Federal University of Technology, Akure, Nigeria

Joseph Ojo Abiola-Falemu
Federal University of Technology, Akure, Nigeria

Isaac Olaniyi Aje
Federal University of Technology, Akure, Nigeria

Wai Yee Betty Chiu
Department of Building Services Engineering, The Hong Kong Polytechnic University, Hong Kong

Fung Fai Ng
Department of Real Estate and Construction, The University of Hong Kong

Derek Henry Thomas Walker
School of Property, Construction and Project Management, RMIT University, Australia

James Harley
School of Property, Construction and Project Management, RMIT University, Australia

Anthony Mills
School of Architecture and Built Environment, Deakin University, Australia

Abukar Warsame
Royal Instititute of Technology (KTH), Sweden

Han-Suck Song
Royal Instititute of Technology (KTH), Sweden

Hans Lind
Royal Instititute of Technology (KTH), Sweden

Ibrahim Mahamid
Hail University, Kingdom of Saudi Arabia